公務員入職

英文運用

第三版

熱門試題王

Fong Sir 著

174條練習題 + 2份模擬試卷 詳盡題解
4大題型熱門試題 密集操練一take過關

序言

未來數年，退休公務員的人數將持續增加，到2023年更是公務員的退休高峰期。每年政府均透過CRE公務員綜合招聘考試來招募人才，為市民提供優質服務。有志加入政府公務員團隊的人士，是時候作好準備，增加獲聘的機會。

公務員綜合招聘考試包括三份45分鐘的試卷，分中文運用、英文運用、基本法測試和能力傾向測試。能力傾向測試全卷共35題選擇題，限時45分鐘。

本書的內容特色如下：

- 收錄174條練習題及2份模擬試卷，全面剖析英文運用考核要訣。

- 深入分析4大題型：Comprehension、Error Idenification、Sentence Completion、Paragraph Improvement的重點概念、解題和答題技巧，讓考生舉一反三，遇到何種題目也應付自如。

- 搜羅熱門試題題型，助你熟習出題方向。

- 試題詳盡講解，方便考生了解自己不足，從而作針對練習。

目錄

目錄

Comprehension

Passage 1 (Question 1-7)

Marie Curie was one of the most accomplished scientists in history. Together with her husband, Pierre, she discovered radium, an element widely used for treating cancer, and studied uranium and other radioactive substances. Pierre and Marie's amicable collaboration later helped to unlock the secrets of the atom.

Marie was born in 1867 in Warsaw, Poland, where her father was a professor of physics. At an early age, she displayed a brilliant mind and a blithe personality. Her great exuberance for learning prompted her to continue with her studies after high school. She became disgruntled, however, when she learned that the university in Warsaw was closed to women. Determined to receive a higher education, she defiantly left Poland and in 1891 entered the Sorbonne, a French university, where she earned her master's degree and doctorate in physics.

Marie was fortunate to have studied at the Sorbonne with some of the greatest scientists of her day, one of whom was Pierre Curie. Marie and Pierre were married in 1895 and spent many productive years working together in the physics laboratory. A short time after they discovered radium, Pierre was killed by a horse-drawn wagon in 1906. Marie was stunned by this horrible misfortune and endured heartbreaking anguish. Despondently she recalled their close relationship and the joy that they had shared in scientific research. The fact that she had two young daughters to raise by herself greatly increased her distress.

PART ONE
Comprehension
Passage

PART TWO

PART THREE

PART FOUR

PART FIVE

Curie's feeling of desolation finally began to fade when she was asked to succeed her husband as a physics professor at the Sorbonne. She was the first woman to be given a professorship at the world-famous university. In 1911 she received the Nobel Prize in chemistry for isolating radium. Although Marie Curie eventually suffered a fatal illness from her long exposure to radium, she never became disillusioned about her work. Regardless of the consequences, she had dedicated herself to science and to revealing the mysteries of the physical world.

1. The Curies' _____ collaboration helped to unlock the secrets of the atom.

A. friendly

B. competitive

C. courteous

D. industrious

E. chemistry

2. Marie had a bright mind and a _____ personality.

A. strong

B. lighthearted

C. humorous

D. strange

E. envious

3. When she learned that she could not attend the university in Warsaw, she felt _____ .

A. hopeless

B. annoyed

C. depressed

D. worried

E. None of the above

4. Marie _____ by leaving Poland and traveling to France to enter the Sorbonne.

A. challenged authority

B. showed intelligence

C. behaved

D. was distressed

E. Answer not available

5. _____ she remembered their joy together.

A. Dejectedly

B. Worried

C. Tearfully

D. Happily

E. Irefully

6. Her _____ began to fade when she returned to the Sorbonne to succeed her husband.

A. misfortune

B. anger

C. wretchedness

D. disappointment

E. ambition

7. Even though she became fatally ill from working with radium, Marie Curie was never _____ .

A. troubled

B. worried

C. disappointed

D. sorrowful

E. disturbed

Answers and Explanations

Passage 1

1. A
「Amicable」意味著友善(friendly)。這並不意味著「competitive(競爭)」(選項B)，即對立的，雄心勃勃的或侵略性的；「courteous(有禮貌的)」(選項C)，即禮貌；「industrious(勤勞)」(選項D)，即辛勤工作；或「chemistry化學」(選項E)：他們的合作是在物理學，而且這段話明確地將他們的合作描述為「友好」(或「友善」)。

2. B
「Blithe」意味著輕鬆(light-hearted)。這並不意味著「strong(強)」(選項A)、「humorous(幽默)」(選項B)或「funny(有趣)」(選項C)、「strange(奇怪)」(選項D)或「envious(嫉妒)」(選項E)。

3. B
「Disgruntled」意味「煩人」(annoyed，選項B)。這並不意味著「hopeless(無望)」(選項A)、「depressed(沮喪)」(選項C)或「worried(擔心)」(選項D)。

4. A
Marie去Sorbonne挑戰權威，因為華沙的大學沒有錄取女生。內文通過描述她「挑釁地」離開波蘭去法國，正好表明了這一挑戰；即她無視權威。這段文字並不表示她表現出「intelligence(智力)」(選項B)、「behaved(表現得好)」(選項C)，或「distressed(感到痛心)」(選項D)，或因她的舉動感到不安。

5. A
「despondently(沮喪地)」的同義詞是「dejectedly(沮喪地)」。文中通過描述Curie的情緒狀態是對丈夫突然意外死亡的「heartbreaking angush(令人心碎的痛苦)」之一。她在文中沒有被描述為受回憶影響而表現出「worried(擔心)」(選項B)，或者「tearfully(含淚)」(選項C)、「happily(愉快)」(選項D)或「irefully(憤怒)」(選項E)。

PART ONE
Comprehension
Passage

PART TWO

PART THREE
Sentence

PART FOUR

PART FIVE

6. C

跟文中所描述的「feeling of desolation(絕望感)」最接近的同義詞是「wretchedness(不幸的)」。

7. C

「Disillusioned(幻滅)」意味著「disappointed(失望)」。它並不意味著「troubled(困擾)」(選項A)、「worried(擔心)」(選項B)或「sorrowful(悲傷)」(D)或「disturbed(受到干擾)」(選項E)。

Passage 2 (Question 8-12)

Mount Vesuvius, a volcano located between the ancient Italian cities of Pompeii and Herculaneum, has received much attention because of its frequent and destructive eruptions. The most famous of these eruptions occurred in A.D. 79.

The volcano had been inactive for centuries. There was little warning of the coming eruption, although one account unearthed by archaeologists says that a hard rain and a strong wind had disturbed the celestial calm during the preceding night. Early the next morning, the volcano poured a huge river of molten rock down upon Herculaneum, completely burying the city and filling the harbor with coagulated lava.

Meanwhile, on the other side of the mountain, cinders, stone and ash rained down on Pompeii. Sparks from the burning ash ignited the combustible rooftops quickly. Large portions of the city were destroyed in the conflagration. Fire, however, was not the only cause of destruction. Poisonous sulfuric gases saturated the air. These heavy gases were not buoyant in the atmosphere and therefore sank toward the earth and suffocated people.

Over the years, excavations of Pompeii and Herculaneum have revealed a great deal about the behavior of the volcano. By analyzing data, much as a zoologist dissects an animal specimen, scientists have concluded that the eruption changed large portions of the area's geography. For instance, it turned the Sarno River from its course and raised the level of the beach along the Bay of Naples. Meteorologists studying these events have also concluded that Vesuvius caused a huge tidal wave that affected the world's climate.

In addition to making these investigations, archaeologists have been able to study the skeletons of victims by using distilled water to wash away the volcanic ash. By strengthening the brittle bones with acrylic paint, scientists have been able to examine the skeletons and draw conclusions about the diet and habits of the residents. Finally, the excavations at both Pompeii and Herculaneum have yielded many examples of classical art, such as jewelry made of bronze, which is an alloy of copper and tin. The eruption of Mount Vesuvius and its tragic consequences have provided everyone with a wealth of data about the effects that volcanoes can have on the surrounding area. Today, volcanologists can locate and predict eruptions, saving lives and preventing the destruction of other cities and cultures.

8. Herculaneum and its harbor were buried under _____ lava.

A. liquid
B. solid
C. flowing
D. gas
E. Answer not available

9. The poisonous gases were not _____ in the air.

A. able to float
B. visible
C. able to evaporate
D. invisible
E. able to condense

10. Scientists analyzed data about Vesuvius in the same way that a zoologist _____ a specimen.

A. describes in detail
B. studies by cutting apart
C. photographs
D. chart
E. Answer not available

公務員入職 **英文運用熱門試題王**

PART ONE
Comprehension
Passage

PART TWO
PART THREE
PART FOUR
PART FIVE

11. _____ have concluded that the volcanic eruption caused a tidal wave.

A. Scientists who study oceans

B. Scientists who study atmospheric conditions

C. Scientists who study ash

D. Scientists who study animal behavior

E. Answer not available in article

12. Scientists have used _____ water to wash away volcanic ash from the skeletons of victims.

A. bottled

B. volcanic

C. purified

D. sea

E. fountain

Answers and Explanations

Passage 2

8. B
「Coagulated」意味著固化(solid)。「Liquid(液體)」(選項A)與固體相反。「Flowing(流動)」(選項C)呈現液態而非固態。「Gas(氣體)」(選項D)是固態的另一個相反。(物質的三種狀態，如火山物質，是液體、固體和氣體。)

9. A
「Buoyant」意味著「能夠漂浮」。內文通過表明氣體因此向地球沉降，並使人窒息而表明了這一點。浮力並不意味著「visible(可見)」(選項B)或可能看到。能漂浮/漂浮並不意味著能夠「evaporate(蒸發)」(選項C)。蒸發意味著轉向蒸氣，只有液體可以。氣體已經是蒸氣。浮力並不意味著「invisible(看不見)」(選項D)或看不見。能夠漂浮並不意味著「able to condense(能夠冷凝)」(選項E)，即從蒸氣轉變為液體。

10. B
「Dissect(解剖)」意味著切割開去進行研究。這並不意味著「describe in detail(詳細描述)」(選項A)、「photographs(拍照)」(選項C)或「chart(繪製)」(選項D)標本。

11. B
氣象學家是研究大氣條件，特別是天氣的科學家。研究「海洋的科學家」(選項A)是海洋學家，即海洋科學家。研究「ash」(選項C)的科學家不是作為一門獨立學科的成員而存在。氣候科學家和許多關注其影響的研究人員研究火山灰。研究「動物行為的科學家」(選項D)是行為主義者或動物行為學家，並且不研究灰燼。

12. C
「Distilled water(蒸餾水)」是純化水。蒸餾水不等於「bottled(瓶裝)」(選項A)、「volcanic(火山)」(選項B)、「sea(海水)」(選項D)或「fountain(噴泉水)」(選項E)。

PART ONE
Comprehension
Passage

PART TWO
Error Identification
Test

PART THREE
Sentence
Completion Test

PART FOUR
Paragraph
Improvement Test

PART FIVE
Mock Paper

Passage 3 (Question 13-17)

One of the most intriguing stories of the Russian Revolution concerns the identity of Anastasia, the youngest daughter of Czar Nicholas II. During his reign over Russia, the czar had planned to revoke many of the harsh laws established by previous czars. Some workers and peasants, however, clamored for more rapid social reform. In 1918, a group of these people known as Bolsheviks overthrew the government. On July 17 or 18, they murdered the czar and what was thought to be his entire family.

Although witnesses vouched that all the members of the czar's family had been executed, there were rumors suggesting that Anastasia had survived. Over the years, a number of women claimed to be Grand Duchess Anastasia. Perhaps the most famous claimant was Anastasia Tschaikovsky, who was also known as Anna Anderson.

In 1920, 18 months after the czar's execution, this terrified young woman was rescued from drowning in a Berlin river. She spent two years in a hospital, where she attempted to reclaim her health and shattered mind. The doctors and nurses thought that she resembled Anastasia and questioned her about her background. She disclaimed any connection with the czar's family. Eight years later, however, she claimed that she was Anastasia. She said that she had been rescued by two Russian soldiers after the czar and the rest of her family had been killed. Two brothers named Tschaikovsky had carried her into Romania. She had married one of the brothers, who had taken her to Berlin and left her there, penniless and without a vocation. Unable to invoke the aid of her mother's family in Germany, she had tried to drown herself.

During the next few years, scores of the czar's relatives, ex-servants, and acquaintances interviewed her. Many of these people said that her looks and mannerisms were evocative of the Anastasia that they had known. Her grandmother and other relatives denied that she was the real Anastasia, however.

Tired of being accused of fraud, Anastasia immigrated to the United States in 1928 and took the name Anna Anderson. She still wished to prove that she was Anastasia, though, and returned to Germany in 1933 to bring suit against her mother's family. There she declaimed to the court, asserting that she was indeed Anastasia and deserved her inheritance.

In 1957, the court decided that it could neither confirm nor deny Anastasia's identity. Although it will probably never be known whether this woman was the Grand Duchess Anastasia, her search to establish her identity has been the subject of numerous books, plays, and movies.

13. Some Russian peasants and workers _____ for social reform.

A. longed

B. cried out

C. begged

D. hoped

E. thought much

14. Witnesses _____ that all members of the czar's family had been executed.

A. gave assurance

B. thought

C. hoped

D. convinced some

E. Answer not available

15. Tschaikovsky initially _____ any connection with the czar's family.

A. denied

B. stopped

C. noted

D. justified

E. Answer not available

16. She was unable to _____ the aid of her relatives.

A. locate

B. speak about

C. call upon

D. identify

E. know

17. In court she _____ maintaining that she was Anastasia and deserved her inheritance.

A. finally appeared

B. spoke forcefully

C. gave testimony

D. gave evidence

E. Answer not available

PART ONE
Comprehension
Passage

PART TWO
Error Identification
Test

PART THREE
Sentence
completion Test

PART FOUR
Paragraph
Improvement Test

PART FIVE
Mock Paper

Answers and Explanations

Passage 3

13. B
「 clamor for 」意味著呼喊(某事)。它並不意味著「longed(要)」(選項A)它，「begged(乞求)」(選項C)、「hoped(希望)」(選項D)，或「 thought much (多想)」(選項E)。

14. A
「vouch」是指保證。它並不意味「thought(著想)」(選項B)、「hoped(希望)「(選項C)或「convinced some(説服某些人)」(選項D)。

15. A
「Disclaimed」是指否認，即拒絕或宣布不真實。它並不意味著「stopped(停止)」(選項B)、「noted(指出)」(選項C)或「justified(證明)」(選項D)，即證實或證實，與被否認的相反。

16. C
她無法援引，即呼籲親屬的援助。調用並不意味著「locate(找到)」(選項A)或找到；「speak about(談論)」(選項B)或討論、「identify(識別)」(選項D)，或「know(知道)」(選項E)。

17. B
「Declaimed(聲明)」意味著有力地説話。這並不意味「finally appeared(最終出現)」(選項A)。儘管她在法庭上也「gave testimony(提供了證詞)」(選項C)，但「聲明」並不意味著作證；它描述了她在這樣做時所説的方式。「宣稱」也並不意味著她「gave evidence(提供了證據)」(選項D)。

Passage 4 (Question 18-19)

King Louis XVI and Queen Marie Antoinette ruled France from 1774 to 1789, a time when the country was fighting bankruptcy. The royal couple did not let France's insecure financial situation limit their immoderate spending, however. Even though the minister of finance repeatedly warned the king and queen against wasting money, they continued to spend great fortunes on their personal pleasure. This lavish spending greatly enraged the people of France. They felt that the royal couple bought its luxurious lifestyle at the poor people's expense.

Marie Antoinette, the beautiful but exceedingly impractical queen, seemed uncaring about her subjects' misery. While French citizens begged for lower taxes, the queen embellished her palace with extravagant works of art. She also surrounded herself with artists, writers, and musicians, who encouraged the queen to spend money even more profusely.

While the queen's favorites glutted themselves on huge feasts at the royal table, many people in France were starving. The French government taxed the citizens outrageously. These high taxes paid for the entertainments the queen and her court so enjoyed. When the minister of finance tried to stop these royal spendthrifts, the queen replaced him. The intense hatred that the people felt for Louis XVI and Marie Antoinette kept building until it led to the French Revolution. During this time of struggle and violence (1789-1799), thousands of aristocrats, as well as the king and queen themselves, lost their lives at the guillotine. Perhaps if Louis XVI and Marie Antoinette had reined in their extravagant spending, the events that rocked France would not have occurred.

18. The people surrounding the queen encouraged her to spend money _____ .

A. wisely
B. abundantly
C. carefully
D. foolishly
E. joyfully

19. The minister of finance tried to curb these royal _____ .

A. aristocrats
B. money wasters
C. enemies
D. individuals
E. spenders

Answers and Explanations

Passage 4

18. B
「Profusely」意味著富有、豐富或過度。它不是「wisely(明智地)」(選項A)或「abundantly(謹慎的)」(選項B)，這兩者在含義上與大量支出的過度內涵相反。「foolishly(愚蠢地)」(選項D)可以與大量的花費相關聯，但不具有相同的含義。大意並不意味著「joyfully(快樂)」(選項E)，即快樂或愉快。

19. B
「Spendthrifts」意味「浪費金錢」，它並不意味著「aristocrats(貴族或特權人士)」(選項A)、「enemies(敵人)」(選項C)和「individuals(個人)」(選項D)。「Spenders(消費者)」(選項E)表示那些花錢的人，但並沒有像「spendthrifts」那樣表達浪費的意思。

PART ONE
Comprehension
Passage

PART TWO
ure Identification
Test

PART THREE
Sentence
completion Test

PART FOUR
Paragraph
Improvement Test

PART FIVE
Mock Paper

Passage 5(Question 20-25)

Many great inventions are initially greeted with ridicule and disbelief. The invention of the airplane was no exception. Although many people who heard about the first powered flight on December 17, 1903 were excited and impressed, others reacted with peals of laughter. The idea of flying an aircraft was repulsive to some people. Such people called Wilbur and Orville Wright, the inventors of the first flying machine, impulsive fools. Negative reactions, however, did not stop the Wrights. Impelled by their desire to succeed, they continued their experiments in aviation.

Orville and Wilbur Wright had always had a compelling interest in aeronautics and mechanics. As young boys they earned money by making and selling kites and mechanical toys. Later, they designed a newspaper-folding machine, built a printing press, and operated a bicycle-repair shop. In 1896, when they read about the death of Otto Lilienthal, the brothers' interest in flight grew into a compulsion.

Lilienthal, a pioneer in hang-gliding, had controlled his gliders by shifting his body in the desired direction. This idea was repellent to the Wright brothers, however, and they searched for more efficient methods to control the balance of airborne vehicles. In 1900 and 1901, the Wrights tested numerous gliders and developed control techniques. The brothers' inability to obtain enough lift power for the gliders almost led them to abandon their efforts.

After further study, the Wright brothers concluded that the published tables of air pressure on curved surfaces must be wrong. They set up a wind tunnel and began a series of experiments with model wings. Because of their efforts, the old tables were repealed in time and replaced by the first reliable figures for air pressure on curved surfaces. This work, in turn, made it possible for the brothers to design a machine that would fly. In 1903 the Wrights built their first airplane, which cost less than $1,000. They even designed and built their own source of propulsion-a lightweight gasoline engine. When they started the engine on December 17, the airplane pulsated wildly before taking off. The plane managed to stay aloft for 12 seconds, however, and it flew 120 feet.

By 1905, the Wrights had perfected the first airplane that could turn, circle, and remain airborne for half an hour at a time. Others had flown in balloons and hang gliders, but the Wright brothers were the first to build a full-size machine that could fly under its own power. As the contributors of one of the most outstanding engineering achievements in history, the Wright brothers are accurately called the fathers of aviation.

20. The idea of flying an aircraft was _____ to some people.

A. boring
B. distasteful
C. exciting
D. needless
E. Answer not available

21. People thought that the Wright brothers had _____ .

A. acted without thinking
B. been negatively influenced
C. been too cautious
D. been mistaken
E. acted in a negative way

22. The Wrights' interest in flight grew into a _____ .

A. financial empire
B. plan
C. need to act
D. foolish thought
E. Answer not available

23. Lilienthal's idea about controlling airborne vehicles was _____ the Wrights.

A. proven wrong by

B. opposite to the ideas of

C. disliked by

D. accepted by

E. improved by

24. The old tables were _____ and replaced by the first reliable figures for air pressure on curved surfaces.

A. destroyed

B. invalidated

C. multiplied

D. approved

E. not used

25. The Wrights designed and built their own source of _____ .

A. force for moving forward

B. force for turning around

C. turning

D. force for going backward

E. None of the above

PART ONE
Comprehension
Passage

PART TWO
Error Identification
Test

PART THREE
Sentence
completion Test

PART FOUR
Paragraph
Improvement Test

PART FIVE
Mock Paper

Answers and Explanations

Passage 5

20. B
「Repulsive」意味著令人厭惡。它並不意味著「boring(無聊)」(選項A)、「exciting(令人興奮)」(選項C)或「needless(不需要)」(選項D)。

21. A
「Impulsive」是指以衝動，即無需思考就行動。人們認為Wrights是「impulsive fools」並不意味著他們認為Wrights「been negatively influenced(受到了負面影響)」(選項B)、「been too cautious(過於謹慎)」(選項C)、「been mistaken(錯誤)」(選項D)或以「acted in a negative way(負面方式行事)」(選項E)。

22. C
「compulsion」是一種行為的需要或衝動。它不是「financial empire(金融帝國)」(選項A)、「plan(計劃)」(選項B)或「foolish thought(愚蠢的思想)」(選項D)。

23. C
「Repellent」意味著「無禮」或「可恨」；換句話說，Lilienthal的想法被Wrights所厭惡，這並不意味著他的想法「opposite to the ideas of Wrights(與Wrights的想法相反)」(選項B)。這意味著它為Wrights所接受(選項D)的相反。他們發現他的想法不愉快，而不是「改善」(選項E)。

24. B
「Repealed」表示「廢除」或「無效」，即未經證實或推翻。它並不意味著被「destroyed(破壞)」(選項A)、「multiplied(乘以)」(選項C)，即「增加/批准」(選項D)反義詞；或「not used(未使用)」(選項E)。

25. A
「Propulsion」是推動或前進的力量。它並不意味著「force for turning around(轉向力量)」(選項B)、「turning(轉向)」(選項C)、「force for going backward(可能振動或後退力量)」(選項D)(如斥力)。

公務員入職 **英文運用**熱門試題王

Error
Identification

II. Error Identification

Knowledge on use of the language is tested through identification of language errors which may be lexical, grammatical or stylistic.

Example:

The sentence below may contain a language error. Identify the part (underlined and lettered) that contains the error or choose 'E No error' where the sentence does not contain an error.

Irrespective for the outcome of the probe, the whole sorry affair has already cast a shadow over this man's hitherto unblemished record as a loyal servant to his country.

A. Irrespective for
B. sorry
C. cast
D. hitherto
E. No error

Answer : A

PART ONE
Comprehension
Passage

PART TWO
Error Identification
Test

PART THREE
Sentence
Completion Test

PART FOUR
Paragraph
Improvement Test

PART FIVE
Mock Paper

Test 1

1. Obesity is an enormous <u>problem,</u> it <u>affects</u> millions of people <u>world-wide</u>, and is an <u>impediment to</u> social life.

A. problem,

B. affects

C. worldwide

D. impediment to

E. No error

2. Mary <u>has taken</u> steps to ensure that <u>she</u> can handle the pressure and anxiety <u>associated with</u> the job, <u>including</u> joining a squash class and enlisting the support of a network of friends.

A. has taken

B. she

C. associated with

D. including

E. No error

3. If you are <u>sure that</u> you are in the <u>right,</u> you <u>would not</u> mind an independent <u>examination of</u> the case.

A. sure that

B. right,

C. would not

D. examination of

E. No error

4. The union <u>insisted</u> on an increase in <u>their</u> <u>members'</u> overtime pay, and threatened to call a strike if the company <u>refused to</u> meet the demand.

A. insisted

B. their

C. members'

D. refused to

E. No error

5. Television viewers claim <u>that</u> the number of scenes <u>depicting</u> child abuse <u>have</u> increased dramatically <u>over</u> the last decade.

A. that

B. depicting

C. have

D. over

E. No error

6. Workers with <u>less</u> personal problems <u>are</u> <u>likely</u> to be <u>more</u> productive.

A. less

B. are

C. likely

D. more

E. No error

PART ONE
Comprehension
Passage

PART TWO
Error Identification
Test

PART THREE
Sentence
Completion Test

PART FOUR
Paragraph
Improvement Test

PART FIVE
Mock Paper

7. The three richest men in America <u>have</u> assets worth more <u>than</u> the <u>combined assets</u> of the fifty poorest countries <u>of</u> the world.

A. have

B. than

C. combined assets

D. of

E. No error

8. <u>Shipwrecked</u> on an uninhabited island, coconuts and <u>other</u> fruits <u>formed</u> the basis of the <u>sailor's</u> diet.

A. Shipwrecked

B. other

C. formed

D. sailor's

E. No error

9. Forty percent of the people alive today <u>have</u> never made a phone call, <u>but</u> thirty percent <u>still</u> have no drainage connections to <u>their</u> homes.

A. have

B. but

C. still

D. their

E. No error

10. The rhododendron, which ornaments so many English gardens, is not native to Europe.

A. , which

B. so many

C. is

D. to

E. No error

11. The fisherman should not have been so careless as to leave the door of the house unbolted when he had gone to work.

A. have been

B. as

C. when

D. had gone

E. No error

12. A census of the island revealed a population of only 20,000 people.

A. A census

B. revealed

C. only

D. people

E. No error

13. The engineer, who is renowned for his <u>ingenuity,</u> <u>has designed</u> a <u>very unique</u> cooling system for our new plant <u>in</u> Germany.

A. ingenuity
B. has designed
C. very unique
D. in
E. No error

14. Shoes of <u>those</u> kind <u>are</u> bad for the feet<u>;</u> low heels <u>are</u> better.

A. those
B. are(the first one)
C. ;
D. are(the second one)
E. No error

15. My dad saw <u>how much</u> Uncle David <u>was enjoying</u> his early retire-ment<u>, and</u> so he decided to <u>do the same</u>.

A. how much
B. was enjoying
C. , and
D. do the same
E. No error

Answers and Explanations

Test 1

1. A
這是「逗號拼接」(Comma splice)的錯誤示例，考生不能使用逗號來連接兩個完整句子。相反，要使用分號(；)。

2. D
錯位修飾詞(Misplaced modifier)。「Including」似乎指的是「job」多於「steps」。

3. C
當「if語句」後使用「現在時態」(present tense)後，我們需要在主句中使用「將來時態」(future tense)。將「would」改為「will」才正確。

4. B
將「their」改為「its」，因「the union」屬單數名詞。

5. C
將「have」改為「has」，因為「the number」需要使用「單數動詞」(singular verb)描述。

6. A
「fewer」適用於形容「可數名詞」(countable)，至於「less」通常用來形容「不可數名詞」(uncountable)。

7. E
這句沒有錯誤。

8. A
這是一個「懸垂修飾語」(dangling modifier)的例子。這意味著椰子遭遇了海難。要改正這個句子，你可以將「Shipwrecked」改為「When he was ship-wrecked」(當他遇難船)時。

9. B
「But」在這裡不適用，改用「and」會較好。

PART ONE
~~Comprehension~~
~~Passage~~

PART TWO
Error Identification
Test

PART THREE
~~Sentence~~
~~Completion Test~~

PART FOUR
~~Paragraph~~
~~Improvement Test~~

PART FIVE
~~Mock Paper~~

10. E
這句沒有錯誤。

11. D
這裡不需要使用「過去完成時態」(past perfect tense)，使用「went」。

12. D
「People」在這裡是多餘的。

13. C
你不能說「very unique」。(就像你不能說「more superior」那樣)

14. A
由於「Kind」是單數(singular)，所以我們應使用「this/that」，而不應使用「these/those」。

15. D
「do the same」的意思較含糊，改為「take early retirement」或類似的字眼會較適合。

Test 2

1. We have no choice <u>but</u> to appoint <u>Apple:</u> she is the <u>best</u> of the two candidates, and there <u>is</u> no prospect of finding more applicants.

A. but
B. Apple:
C. best
D. is
E. No error

2. The reason I <u>will</u> not be <u>going</u> to Brazil this year is <u>because</u> I will use up all my travel money <u>attending</u> an important conference in Thailand.

A. will
B. going
C. because
D. attending
E. No error

3. If you <u>were</u> to work at least three hours a day on the project, we <u>would</u> complete it in a <u>shorter</u> time, and with <u>less</u> problems.

A. were
B. would
C. shorter
D. less
E. No error

PART ONE
Comprehension Passage

PART TWO
Error Identification Test

PART THREE
Sentence completion Test

PART FOUR
Paragraph improvement test

PART FIVE
Mock Paper

4. The manager tried hard to <u>effect</u> a change in company policy<u>, but</u> the owner, who steadfastly refused <u>to compromise,</u> <u>overruled</u> him on every point.

A. effect

B. , but

C. to compromise

D. overruled

E. No error

5. The new cinema is <u>undoubtedly</u> well stocked and functional<u>, but</u> no one can say that <u>its</u> atmosphere is anything like <u>the old one</u>.

A. undoubtedly

B. , but

C. its

D. the old one

E. No error

6. My uncle, <u>who was</u> on leave, along with my two sons and <u>I,</u> <u>went</u> fishing down by the river.

A. who

B. was

C. I

D. went

E. No error

7. <u>Hopefully</u>, we <u>will be</u> able to complete the work <u>before</u> the rainy season sets <u>in</u>.

A. Hopefully

B. will be

C. before

D. in

E. No error

8. You <u>would</u> have to choose <u>her</u>, if you are looking <u>for</u> the best athlete <u>to represent</u> the school.

A. would

B. her

C. for

D. to represent

E. No error

9. All the trapped miners <u>began</u> to <u>lose</u> hope<u>, it</u> had been twenty four hours <u>since</u> the tunnel collapsed.

A. began

B. lose

C. , it

D. since

E. No error

PART ONE
Comprehension
Passage

PART TWO
Error Identification
Test

PART THREE
Sentence
Completion Test

PART FOUR
Paragraph
Improvement Test

PART FIVE
Mock Paper

10. <u>Because</u> they played <u>by</u> the rules, the members of the team <u>were given</u> a standing ovation even though <u>it</u> did not win the match.

A. Because

B. by

C. were given

D. it

E. No error

11. Her <u>avaricious</u> relatives assembled at the <u>lawyer's</u> office to hear <u>the reading of</u> <u>Jemima's will</u>.

A. avaricious

B. lawyer's

C. the reading of

D. Jemima's will

E. No error

12. He was <u>not merely</u> expected to contribute funds to the project, <u>but</u> to work <u>as</u> hard as the <u>other</u> patrons.

A. not merely

B. but

C. as

D. other

E. No error

13. None of us knows what the outcome of the battle between the coordinator and us will be.

A. of us

B. knows

C. between

D. us

E. No error

14. Neither of my brothers do anything to make life better for our parents who are both suffering from arthritis.

A. do

B. our

C. who

D. from

E. No error

15. The teacher sat down besides the frightened child and tried to reassure him that the monster was merely imaginary.

A. sat

B. besides

C. to reassure

D. merely

E. No error

PART ONE
Comprehensive Essays

PART TWO
Error Identification Test

PART THREE
Sentence Completion Test

PART FOUR
Paragraph Improvement Test

PART FIVE
Mark Paper

Answers and Explanations

Test 2

1. C
因為題目內只有兩名候選人，所以只需使用「better」(較好)而不是「best」(最好)。

2. C
正確的用法是說「the reason is that」(原因是)，而不是「the reason is because」(原因是因為)：寫得「原因」，就不需要寫「因為」。

3. D
將「less」改為「fewer」。「fewer」適用於形容「可數名詞」(countable)，「less」通常用來形容「不可數名詞」(uncountable)。

4. E
這句沒有錯誤。

5. D
不正確的比較(Incorrect comparison)。「氣氛」(Atmosphere)必須與「氣氛」相比。我們可以寫「...its atmosphere is anything like that of the old one」。

6. C
將「I」更改為「me」(介詞的對象)。

7. A
「Hopefully」被錯誤地用作修飾名詞「we」的形容詞。「Hopefully」實際上是一個副詞(adverb)，所以應該用來修飾一個動詞或另一個副詞。
正確的寫法是：「With hope, we will be able to complete the work before the rainy season sets in.」

8. A
將「would」更改為「will」。

9. C
因不能用逗號(comma)連接兩個句子，故要將逗號更改為分號(semi-colon)。

10. D
因為「It」沒有「先行詞」(antecedent)，故將「It」改為「they/the team」。

11. E
這句沒有錯誤。

12. A
為了使句子結構平行，將「not merely」放在「expected」之後。

13. E
這句沒有錯誤。

14. A
由於「Neither」是單數，所以要將「do」改為「does」。

15. B
將「besides」改為「beside」。

Test 3

1. A number of workers <u>who</u> take this course every year <u>find</u> that <u>their</u> knowledge of English <u>is</u> inadequate.

A. who

B. find

C. their

D. is

E. No error

2. Either of the solutions you <u>have</u> suggested <u>are</u> acceptable to the committee, <u>whose</u> members <u>are willing</u> to compromise.

A. have

B. are

C. whose

D. are willing

E. No error

3. The last man on earth <u>will</u> abandon his ruined house <u>for</u> a cave, <u>and</u> his woven clothes for an <u>animal's</u> skin.

A. will

B. for

C. , and

D. animal's

E. No error

4. The station was a <u>hive</u> of bustling <u>activity,</u> the arrival of the train was the <u>most important</u> event of the day <u>in</u> that remote place.

A. hive

B. activity,

C. most important

D. in

E. No error

5. Tom's grandfather's legacy <u>is</u> substantial, <u>especially</u> if the value of the rare stamps <u>are</u> taken <u>into</u> consideration.

A. is

B. especially

C. are

D. into

E. No error

6. <u>Either</u> you or I <u>are</u> <u>going</u> <u>to do</u> the washing-up.

A. Either

B. are

C. going

D. to do

E. No error

7. The bridal gown was <u>most</u> unique<u>: the</u> bridegroom designed <u>it</u> and <u>his</u> mother provided the lace fabric.

A. most
B. : the
C. it
D. his
E. No error

8. For a successful career <u>as</u> a beautician, <u>one</u> must be prepared to <u>dissemble:</u> you must not tell your client the unvarnished truth about <u>his or her</u> appearance.

A. as
B. one
C. dissemble:
D. his or her
E. No error

9. When Russell Wallace and Darwin <u>independently</u> proposed similar theories, Darwin <u>had</u> already accumulated extensive evidence <u>with which</u> to support <u>his</u> ideas.

A. independently
B. had
C. with which
D. his
E. No error

10. Everyone <u>who</u> visits Hong Kong <u>is impressed</u> by its cleanliness, <u>which</u> is mainly a result of rigorous implementation of <u>their</u> strict laws.

A. who

B. is impressed

C. which

D. their

E. No error

11. Ann wondered <u>whether</u> the city had changed <u>alot</u> since she had left to <u>go to</u> study abroad<u>.</u>

A. whether

B. alot

C. go to

D. .

E. No error

12. The company <u>bowed</u> to <u>pressure,</u> now <u>it</u> has removed the offensive advertisement <u>from</u> the hoarding.

A. bowed

B. pressure,

C. it

D. from

E. No error

13. I <u>will</u> not object to <u>his</u> delivering the lecture <u>as</u> long as he is told not to make personal attacks <u>on</u> his critics.

A. will

B. his

C. as

D. on

E. No error

14. <u>While</u> he thinks the <u>phenomenon</u> is the result of enzyme action, I believe it is <u>caused by</u> a shortage <u>of</u> a neurotransmitter.

A. While

B. phenomenon

C. caused by

D. of

E. No error

15. Mary argued vehemently <u>with</u> her mother <u>over</u> <u>her</u> refusal <u>to at-tend</u> the Christmas ball.

A. with

B. over

C. her

D. to attend

E. No error

Answers and Explanations

Test 3

1. E
這句沒有錯誤。

2. B
因為「Either」需要一個單數動詞（singular verb），所以需要將「are」改為「is」。

3. E
這句沒有錯誤。

4. B
將comma(逗號)改為分號(semi colon)，因前、後兩句是獨立的句子。

5. C
「are」應改為「is」以作回應「value」一詞。

6. B
當使用「either...or...」或「neither...nor...」作為句子的開頭，有一點我們必須注意：
「either (subject 1) or (subject 2)」或「neither (subject 1) nor (subject 2)」隨後的動詞必須與「subject 2」相符。我們稱這規則為「principle of proximity」。「Proximity」就是「相近」的意思。因為「subject 2」與隨後的動詞較「subject 1」相近，所以動詞與「subject 2」相應。正確的寫法應為：Either you or I am going to do the washing-up.

7. A
永遠不要在「unique」前放置「most」(最)。(因「獨特」經已是最高級了)

8. B
對於「代詞連續性」(pronoun continuity)，將「one」改為「you」。

9. D
代詞語意含糊。證據是支持Darwin或Wallace，或兩者？

PART ONE
Comprehension
Passage

PART TWO
Error Identification
Test

PART THREE
Sentence
completion Test

PART FOUR
Paragraph
Improvement Test

PART FIVE
Mock Paper

10. D
當談論一個國家或城市時，你不能使用「they」或「their」。

11. B
「a lot」應分為兩個字。

12. B
逗號拼接錯誤。正確答案是要將「壓力」一詞後面的標點符號，由逗號改為分號。

13. E
這句沒有錯誤。

14. A
「While」應用於表達「同時採取行動」的意思，正確答案要改為「though」。

15. C
代詞語意含糊：「her」是Mary還是她的母親？

Test 4

1. I agree <u>that</u> a knowledge of Latin is helpful <u>to build</u> a good English vocabulary<u>, but</u> I do not think I have the capacity <u>to</u> study the subject at the moment.

A. that

B. to build

C. , but

D. to

E. No error

2. In <u>such</u> areas as sports, ranking of individual performance <u>is</u> relatively well accepted <u>since</u> the parameters on which the rating <u>are</u> based are generally objective.

A. such

B. is

C. since

D. are

E. No error

3. <u>Determination of</u> the long-term <u>effects of</u> aerosols on the upper atmosphere <u>is</u> currently one of the <u>more challenging</u> problems in climate research.

A. Determination of

B. effects of

C. is

D. more challenging

E. No error

4. The <u>most</u> important skill I <u>had learned</u> in my three years <u>of</u> high school was <u>to</u> direct the course of my own studies.

A. most

B. had learned

C. of

D. to

E. No error

5. Scientific advances over the last forty years <u>have led</u> to revolutionary changes in science, agriculture and communication<u>, and</u> generally <u>enhancing</u> socio-econornic development and the quality of our <u>lives</u>.

A. have led

B. , and

C. enhancing

D. lives

E. No error

6. This detailed <u>yet</u> readable biography is well researched <u>and</u> provides valuable insight <u>to</u> the facts <u>that</u> motivated the famous philosopher.

A. yet

B. and

C. to

D. that

E. No error

7. I have <u>nearly</u> written all the new tests <u>for</u> inclusion in the revised edition of my book, and <u>hope to finish</u> the work <u>within</u> a week.

A. nearly

B. for

C. hope to finish

D. within

E. No error

8. The series of letters that Margaret wrote to her father <u>contains</u> a valuable <u>commentary on</u> the prevailing social conditions and attitudes that lead to <u>her</u> leaving home at <u>such</u> a young age.

A. contains

B. commentary on

C. her

D. such

E. No error

PART ONE
Comprehension

PART TWO
Error Identification
Test

PART THREE
Sentence

PART FOUR
Paragraph
Improvement Test

PART FIVE

9. The unfortunate accident <u>that caused</u> the explosion <u>was</u> <u>extensively</u> reported in all the local newspapers and <u>national</u> television.

A. that caused

B. was

C. extensively

D. national

E. No error

10. Neither of the answers provided in the <u>memorandum</u> <u>address</u> my concerns <u>about</u> the <u>validity of</u> the procedure.

A. memorandum

B. address

C. about

D. validity of

E. No error

11. Katz claimed that reading classic novels <u>is</u> more <u>illuminating</u> than <u>to read</u> autobiographies <u>written by</u> their authors.

A. is

B. illuminating

C. to read

D. written by

E. No error

12. The students <u>have been</u> practicing for the concert <u>since</u> three weeks<u>, and</u> in that time <u>have</u> improved considerably.

A. have been

B. since

C. , and

D. have

E. No error

13. Suzanne <u>recounted</u> her <u>improbable</u> tale <u>with</u> enthusiasm and <u>in a convincing manner</u>.

A. recounted

B. improbable

C. with

D. in a convincing manner

E. No error

14. If you <u>were willing</u> to ask for directions, instead of <u>doggedly</u> driving on, we might get to <u>our</u> destination <u>sooner</u>.

A. were willing

B. doggedly

C. our

D. sooner

E. No error

15. <u>Waiting</u> for the results <u>of</u> the final examination, the student's nerves <u>were</u> on edge; she could not sleep properly <u>or</u> eat normally.

A. Waiting
B. of
C. were
D. or
E. No error

Answers and Explanations

Test 4

1. B
正確的寫法是「is helpful in building」或「helps to build」。

2. D
將「are」改為「is」以呼應「rating」。

3. D
將「more challenging」改為「most challenging」。(如只有兩個問題,便使用「more challenging」,但科學家所面對的氣候問題,肯定多於兩個,故本題用「most」。)

4. B
動詞時態錯誤:通過刪除「had」,便能將「過去完成時式」改變為「過去時式」。

5. C
為了呼應句子的結構,我們必須説「have enhanced」。

6. C
本句屬西方諺語:insight into。

7. A
放錯位置:將「nearly」放在「all」前面。

8. E
這句沒有錯誤。

9. D
句中省略了一個重要的詞彙:「national」前面要放上「on」。

10. B
由於「Neither」是單數,故需要一個單數動詞 (singular verb) 配合使用,所以「address」要改為「addresses」。

PART ONE
Comprehension
Passage

PART TWO
Error Identification
Test

PART THREE
Sentence
completion Test

PART FOUR
Paragraph
Improvement Test

PART FIVE
Mock Paper

11. C
為使句子結構平行，將「to read」改為「reading」。

12. B
由於「Since」是用以指出一個明確的時間，而不是泛指一段時間，故將「since」改為「for」。

13. D
由「and」串連起來的兩個項目需要顧及句子的平行結構，所以「in a convicting manner」要改為「conviction」。

14. E
這句沒有錯誤。

15. A
懸掛修飾。因神經(nerves)沒有在等待！所以要將「Waiting」改寫為「As she waited」。

PART
THREE

Sentence Completion

III. Sentence Completion

In this section, candidates are required to fill in the blanks with the best options given. The questions focus on grammatical use.

Example :

Complete the following sentence by choosing the best answer from the options given.

This market research company claims to predict in advance _____ by conducting exit polls of selected voters.

A. the results of an election will be

B. the results will be of an election

C. what results will be of an election

D. what the results of an election will be

E. what will the results of an election be

Answer : D

PART ONE
Comprehension
Passage

PART TWO
Error Identification
Test

PART THREE
Sentence
completion Test

PART FOUR
Paragraph
Improvement Test

PART FIVE
Mock Paper

Test 1

1. Today Wegener's theory is _____ ; however, he died an outsider treated with _____ by the scientific establishment.

A. unsupported – approval

B. dismissed - contempt

C. accepted – approbation

D. unchallenged – disdain

E. unrivalled - reverence

2. The revolution in art has not lost its steam; it _____ on as fiercely as ever.

A. trudges

B. meanders

C. edges

D. ambles

E. rages

3. Each occupation has its own _____ ; bankers, lawyers and computer professionals, for example, all use among themselves language which outsiders have difficulty following.

A. merits

B. disadvantages

C. rewards

D. jargon

E. problems

4. _____ by nature, Jones spoke very little even to his own family members.

A. garrulous

B. equivocal

C. taciturn

D. arrogant

E. gregarious

5. Biological clocks are of such _____ adaptive value to living organisms, that we would expect most organisms to _____ them.

A. clear - avoid

B. meager - evolve

C. significant - eschew

D. obvious - possess

E. ambivalent - develop

6. The peasants were the least _____ of all people, bound by tradition and _____ by superstitions.

A. free - fettered

B. enfranchised - rejected

C. enthralled - tied

D. pinioned - limited

E. conventional - encumbered

7. Many people at that time believed that spices help preserve food; however, Hall found that many marketed spices were _____ bacteria, moulds and yeasts.

A. devoid of

B. teeming with

C. improved by

D. destroyed by

E. active against

8. If there is nothing to absorb the energy of sound waves, they travel on _____ , but their intensity _____ as they travel further from their source.

A. erratically - mitigates

B. eternally - alleviates

C. forever - increases

D. steadily - stabilizes

E. indefinitely - diminishes

9. The two artists differed markedly in their temperaments; Palmer was reserved and courteous, Frazer _____ and boastful.

A. phlegmatic

B. choleric

C. constrained

D. tractable

E. stoic

10. The intellectual flexibility inherent in a multicultural nation has been _____ in classrooms where emphasis on British-American literature has not reflected the cultural _____ of our country.

A. eradicated - unanimity

B. encouraged - aspirations

C. stifled - diversity

D. thwarted - uniformity

E. inculcated - divide

11. The conclusion of his argument, while _____, is far from _____ .

A. stimulating - interesting

B. worthwhile - valueless

C. esoteric - obscure

D. germane - relevant

E. abstruse - incomprehensible

12. In the Middle Ages, the _____ of the great cathedrals did not enter into the architects' plans; almost invariably a cathedral was positioned haphazardly in _____ surroundings.

A. situation - incongruous

B. location - apt

C. ambience - salubrious

D. durability - convenient

E. majesty - grandiose

PART ONE
Comprehension
Passage

PART TWO
Error Identification
Test

PART THREE
Sentence
completion Test

PART FOUR
Paragraph
Improvement Test

PART FIVE
Mock Paper

Answers and Explanations

Test 1

1. D

由於「however」是表示矛盾的陳述，所以要在句子的兩半中有相反的想法。所以，Wegener的理論並沒有受到挑戰(每個人都接受它)。然而他卻被disdain(蔑視)。

此外，句子中的「outsider(局外人)」一詞表示，第二條橫線上必須是否定詞。

(approbation=認同；reverence=尊重)

2. E

句尾的「as ever」表示事物沒有改變，因此句子的兩半需要說相似的事情。

所以，如果「revolution(革命)」沒有失去動力(not lost its steam)，它將會像以前一樣強勁，因此「rages(憤怒)」是最好的詞。

此外，「fiercely(凶狠)」一詞表明我們需要一個強大的詞彙。

(trudges =慢慢走路；meanders =漫步；ambles =漫無目的地行走)

3. D

句子的後半部分正在談論「language(語言)」，它正在擴大在semi-colon(分號)之前所說的話。因此，第一條橫線上需要關於語言，而「jargon(專業術語)」是專門人才使用的專業用語。

4. C

由於逗號後面寫Jones「spoke very little(說得很少)」，因此「taciturn(沉默)」是最好的答案。

(garrulous=健談的；equivocal=模糊的；gregarious=可交際的)

5. D

從句子結構表明，如果「Biological clocks(生物鐘)」具有很大優勢，那麼大多數生物體都擁有它們。(或者，如果它們不是優勢，那麼生物體就不會擁有它們。)

因此，「obvious (顯而易見)」的價值，使我們期望有機體擁有(possess)它們。

(meager=小、輕微；eschew=避免；ambivalent=模棱兩可)

6. A

逗號後面的部分擴展了所述內容。還要注意「least(最少)」一詞，這裡的意思是「not(不)」。

因此，農民是最不自由的，因為他們被迷信束縛「restricted or bound(限制或束縛)」。

(enfranchised=給予投票權；enthralled=著迷；pinioned=被捆綁；encumbered=負擔)

7. B

「However」表示句子的第一部分與第二部分矛盾。

因此，如果人們曾經認為香料保存了食物，那麼句中提到的人發現香料不能保存食物，而且事實上香料中是「teeming(充滿)」了許多可能破壞食物的細菌等。

8. E

由於「but」表明有矛盾。所以，如果波浪沒有被吸收，他們會無限期地旅行顯然是永遠，但當它們離開時，強度變小(減小)。

請記住，該句子必須具有最好的意義，因此說聲音強度隨著波浪傳播而增加與常識相反。

(erratically=不正常的；alleviates =不那麼嚴重)

9. B

既然兩個藝術家「differ」，我們需要對立面。

所以，由於「reserved」和「courteous」都是正面的用詞，我們需要一個意思帶「bad」字的答案放在橫線上。Choleric means(膽小的手段)很容易被激怒，所以會與克制相反。

(phlegmatic=冷靜、沉著冷靜；constrained=克制；tractable=容易引導、溫順；stoic=有毅力)

10. C

嘗試理解這裡的整體含義。強調一種類型的文學(British-American)並沒有體現出作者本國的多元文化差異的「多樣性」(diversity)，所以我們文化的靈活性(flexibility)已經被削弱或壓制、扼殺(stifled)。

(eradicated=消滅；unanimity=共識、協議；agreement=希望；thwarted=阻止；inculcate=灌輸)

PART ONE
Comprehension
Passage

PART TWO
Error Identification
Test

PART THREE
Sentence
completion Test

PART FOUR
Paragraph
Improvement Test

PART FIVE
Mock Paper

11. E

由於「far from」表示需要一個相反的。

所以，雖然結論是abstruse(模糊、難於理解)，但它並不完全incomprehensible(不可理解)。

(esoteric=隱晦；germane=相關)

12. A

使用分號表示句子的第二部分在第一部分擴展。

所以，第二部分告訴我們，我們正在談論大教堂(cathedrals)的位置或情況。由於第一部分告訴我們，建築師沒有註意到情況，大教堂被隨機定位在奇怪的、不協調(incongruous)的環境中。

(incongruous=不匹配、奇怪；apt=適合；ambience=氣氛、環境；grandiose=宏大)

Test 2

1. Unwilling to admit that they had been in error, the researchers tried to _____ their case with more data obtained from dubious sources.

A. ascertain

B. buttress

C. refute

D. absolve

E. dispute

2. Archaeology is a poor profession; only _____ sums are available for excavating sites and even more _____ amounts for preserving the excavations.

A. paltry - meager

B. miniscule - substantial

C. average - augmented

D. judicious - penurious

E. modest - generous

3. The student was extremely foolhardy; he had the _____ to question the senior professor's judgment.

A. wisdom

B. temerity

C. interest

D. trepidation

E. condescension

PART ONE
Comprehension Passage

PART TWO
Error Identification Test

PART THREE
Sentence completion Test

PART FOUR
Paragraph Improvement Test

PART FIVE
Mock Paper

4. The formerly _____ waters of the lake have been polluted so that the fish are no longer visible from the surface.

A. muddy

B. tranquil

C. stagnant

D. pellucid

E. rancid

5. After the accident, the nerves to her arm were damaged and so the muscles _____ through disuse.

A. atrophied

B. contracted

C. elongated

D. invigorated

E. dwindled

6. Some critics maintain that Tennyson's poetry is uneven, ranging from the _____ to the _____ .

A. sublime - elevated

B. trite - inspired

C. vacuous - inane

D. succinct - laconic

E. sonorous - voluble

7. After grafting there is a _____ of lymphocytes in the lymph glands; the newly produced lymphocytes then move in to attack the foreign tissue.

A. diminution

B. proliferation

C. obliteration

D. paucity

E. attraction

8. One _____ the new scheme is that it might actually _____ just those applicants that it was intended to encourage.

A. highlight of - stimulate

B. feature of - attract

C. problem with - induce

D. attraction of - intimidate

E. drawback of - daunt

9. Corruption is _____ in our society; the integrity of even senior officials is _____ .

A. growing - unquestioned

B. endangered - disputed

C. pervasive - intact

D. rare - corrupted

E. rife – suspect

10. In their day to day decision making, many senior managers do not follow the rational model favored by orthodox management experts, but rather rely on intuitive processes that often appear _____ and _____ .

A. cerebral - considered

B. heretical - judgmental

C. conscientious - logical

D. irrational - iconoclastic

E. capricious - deliberate

11. His characteristically _____ views on examination methods at university level have aroused _____ in those who want to introduce innovative and flexible patterns of assessment.

A. hidebound - antagonism

B. moderate - anger

C. reactionary - admiration

D. rigid - support

E. accommodating - annoyance

12. Our grandfather was an entertaining _____ ; he used to _____ us with marvelous anecdotes that we, in our childlike simplicity, accepted unquestioningly.

A. rascal - bore

B. orator - intimidate

C. raconteur - regale

D. curmudgeon - surprise

E. tyrant - stupefy

Answers and Explanations

Test 2

1. B
研究人員不願承認他們犯了錯誤，因此他們會試圖支持(buttress)他們的論點。
(ascertain=發現；refute=證明錯誤；absolve=原諒)

2. A
分號後的部分擴展到句子的第一部分。
因此，由於第一部分告訴我們archaeology中沒有錢，那麼挖掘的數量只會很小、微不足道(paltry)。
另外「 even more 」表示需要另一個類似的詞。因此，微薄(meager)也意味著小。
(miniscule=微小；augmented=增加；judicious=明智、正義；penurious=窮人)

3. B
分號表示第二部分與第一部分密切相關。
所以，既然這個學生是魯莽的(foolhardy)，他就是在做一些不明智的事情。對教授的判斷提出質疑會是魯莽的，所以我們可以説他有過於膽怯(temerity)來質疑他。
請注意，temerity有一個負面的內涵。
(trepidation=恐懼、猶豫；condescension=傲慢)

4. D
「formerly」這個詞表明，一旦事情不同。
所以，從現在開始，水域受到污染，以至於看不到魚，那麼以前它們一定是未受污染的，並且是清澈的(pellucid)。
(tranquil=寧靜；stagnant=不動；pellucid=透明、清澈；rancid=陳舊)

5. A
這句話表明肌肉沒有使用，所以我們希望他們浪費掉。「atrophied」就是意味著浪費、枯萎。
請注意，雖然「dwindled」都有「減少」的意思，但不能用於肌肉上。
(invigorated=激勵)

PART ONE
Comprehension
Passage

PART TWO
Error Identification
Test

PART THREE
Sentence
completion Test

PART FOUR
Paragraph
Improvement Test

PART FIVE
Mock Paper

6. B

由於「Ranging from something to something」表明極端是必需的。我們也被告知詩歌不平衡,也表明需要對立面。

因此,trite(陳腐)是一個消極的詞,而inspired(啟發)是一個積極的詞彙。

(sublime=鼓舞人心;vacuous　無意義=簡潔;succinct=短而重要;laconic=少用詞;sonorous——充滿聲音;voluble=多言)

7. B

分號表示句子的第二部分放大第一部分。第二部分也指出淋巴細胞(lympho-cytes)是「newly produced(新產生的)」。因此,我們推斷腺體中存在這些細胞的生產(proliferation)。

(diminution=減少;proliferation=增長和增殖;obliteration=消除;paucity=不足)

8. E

嘗試理解句子的邏輯。新計劃實際上可能會採取一些措施來「推遲」(daunt),它打算鼓勵的申請人。「intimidate(威嚇)」這個詞也適用於第二條橫線上,但它選項D的「attraction of」不可能是正確的……因為一個具有消極特徵的積極詞彙是不可能的。

(stimulate=鼓勵=誘導;daunt=恐嚇=推遲)

9. E

分號表示第二部分擴展到第一部分。所以,如果腐敗是普遍(rife)的,那麼我們會懷疑官員的完整性。他們的誠信將受到suspect(遭到懷疑)。

(pervasive=遍布到處;rife=普及、普遍)

10. D

「but rather」結構表明相反。由於他們不遵循一個理性模型(rational model),我們推斷他們似乎不合理(irrational)。此外,由於他們不遵循正統(orthodox),他們必須是非傳統的(iconoclastic)。

請注意,「and」通常會鏈接具有相似值(同是正面或負面等)的單詞。

(cerebral=關心思想;heretical=反對正統=反傳統;反复無常capricious=異想天開、善變)

11. A

遵循邏輯。如果他的觀點是靈活的，那麼想要靈活方法的人會批准。但如果他的觀點僵硬，那麼同樣的人會反對他們。因此，最合適的方式是hidebound(僵化)和antagonism(敵意)。

(reactionary=超保守；accommodating=靈活)

12. C

祖父曾經講故事(軼事)。他也很有趣。所以，最好的答案就是他是一個raconteur(說故事的人)，他們regale(愛娛樂)孩子們。

(orator=好的演講者；curmudgeon=暴躁的人；tyrant=苛刻的統治者)

PART ONE
Comprehension Passage

PART TWO
Error Identification Test

PART THREE
Sentence completion Test

PART FOUR
Paragraph Improvement Test

PART FIVE
Mock Paper

Test 3

1. He was treated like a _____ and cast out from his community.

A. ascetic
B. prodigy
C. prodigal
D. pariah
E. tyro

2. The teacher accused me of _____ because my essay was so similar to that of another student.

A. procrastination
B. plagiarism
C. celerity
D. confusion
E. decorum

3. We live in a _____ age; everyone thinks that maximizing pleasure is the point of life.

A. ubiquitous
B. propitious
C. sporadic
D. corrupt
E. hedonistic

4. Thankfully the disease has gone into _____; it may not recur for many years.

A. treatment
B. sequestration
C. quarantine
D. remission
E. oblivion

5. People from all over the world are sent by their doctors to breathe the pure, _____ air in this mountain region.

A. invigorating
B. soporific
C. debilitating
D. insalubrious
E. aromatic

6. As were many colonial administrators, Gregory was _____ in his knowledge of the grammar of the local language, though his accent was _____ .

A. deficient - poor
B. competent - adequate
C. faultless - awful
D. well-versed - effective
E. erratic - eccentric

7. Though Adam Bede is presented to us by the author as _____ fiction, there are none of the life-like meanderings of the story of Amos Barton.

A. realistic
B. romantic
C. imaginative
D. educational
E. entertaining

8. There is a general _____ in the United States that our ethics are declining and that our moral standards are _____ .

A. feeling - normalizing
B. idea - futile
C. optimism - improving
D. complaint - deteriorating
E. outlook - escalating

9. Homo sapiens, the proud splitter of the atom, inventor of the electronic computer, _____ of the genetic code may be humbled by a lowly _____ of the sewers and soils - the microbe.

A. designer - inhabitant
B. discoverer - rodent
C. writer - organism
D. decipherer - denizen
E. author - purifier

10. After centuries of obscurity, this philosopher's thesis is enjoying a surprising _____ .

A. dismissal
B. remission
C. decimation
D. longevity
E. renaissance

11. The threat of war, far from _____ , lay heavily in the air, and the villagers, while _____ going about their normal activities, were unable to shake off the feeling of impending catastrophe.

A. receding - ostensibly
B. diminishing - contentedly
C. increasing - apparently
D. escalating - joyfully
E. subsiding - felicitously

12. Although alarmed by the _____ , Professor Symes had no reason to doubt the _____ of his student's results, for this student was nothing if not reliable.

A. conclusions - folly
B. deductions - impudence
C. implications - veracity
D. errors - truth
E. inferences - invalidity

PART ONE
Comprehension
Passage

PART TWO
Error Identification
Test

PART THREE
Sentence
completion Test

PART FOUR
Paragraph
Improvement Test

PART FIVE
Block Page

Answers and Explanations

Test 3

1. D
從文中提到的「He」被「cast out from his community」，可以知道該人被趕走了。被拋棄是一個賤民。
(ascetic=苦行憎；prodigy=天才，或非常有才華的人；prodigal=浪費的人；tyro=新手，初學者)

2. B
「because」給出了需要的單詞之理由。如果文章如此相似，看起來像是被複製就是plagiarism(抄襲)。
(procrastination=拖延、延遲；celerity=速度；decorum=良好和正確的行為)

3. E
分號後的部分解釋了我們正在談論的年齡。
所以，既然我們被告知最大化快樂就是關鍵，我們需要的是hedonistic(享樂主義)。
(ubiquitous=無處不在；propitious=有利；sporadic=間歇、不連續)

4. D
分號後面的部分解釋了句子的第一部分。
所以，一段時間內可能不會重現的事情將會得到remission(緩解)。
(sequestration=隔離；quarantine=隔離；remission=疾病的暫時改善；oblivion=沒有意識到的狀態)

5. A
由於空氣被描述為「pure」，我們需要一個意義帶正面的詞語。另外，由於醫生推薦它，空氣必須對健康有益。
因此，我們選擇「invigorating」意味著「激勵」。
(soporific=誘導睡眠；debilitating=弱化；insalubrious=不健康；aromatic=令人愉快的氣味)

6. C
「though」表示需要對立面。
因此，我們說他的語法是「faultless(完美)」的，儘管他的口音很「awful(糟糕)」。(這是唯的對立面)
(erratic=不可預測；eccentric=奇數)

7. A

「Though」表示在句子的兩半中需要對立面。此外，「life-like」表示「realis-tic(現實)」是本句需要的詞語。

這句話意味著，儘管沒有像生活一樣的蜿蜒曲折，但作品被呈現為逼真的小說。(請注意：為了從句子中獲得意義，有時候可以幫助轉換部分。)

8. D

「and」通常加入類似含義或重量的東西。這表明由於道德標準下降，道德標準也在deteriorating(下降、惡化)。除了「optimism(樂觀主義)」之外，幾乎所有的詞都適合放在第一條橫線之上。

(futile=徒勞無用、無效；escalating=升級、增加)

9. D

第一條橫線需要傳達人類對遺傳密碼所做的事情：唯一合適的兩個詞是「dis-coverer」或「decipherer」，但由於微生物不是囓齒動物(rodent)，所以我們可以消除這一對。

(rodent=囓齒動物，即像老鼠或老鼠那樣的動物；decipherer=解碼的人；denizen=居民)

10. E

句子告訴我們，論文一直處於obscurity(默默無聞的狀態，容易被遺忘或忽略)，但現在它正在復興。我們可以說它正在經歷renaissance。

(remission=暫時停止疾病；decimation=破壞；longevity=生命的長度)

11. A

「Far from」表示正在進行相反的處理。因此，由於即將發生災難，戰爭的威脅遠未減少。這表明「receding(後退), diminishing(減少), or subsiding(下降)」可能是合適的。我們可以消除B和E，因為兩個選項的搭檔詞「contentedly(滿足地)」和「felicitously(愉快地)」都不適合描述危險感，所以村民顯然(ostensi-bly，表面上)表現正常。

(escalating=增加)

12. C

「Nothing if not」意思是「very」。因此，如果學生非常可靠，教授不會懷疑結果的「truth」或「veracity(真實性)」。由於學生是可靠的，我們可以消除「er-ror」選擇，並選擇「implications」。

(folly=愚蠢；impudence=狡猾；inferences=我們可以推論出的東西)

Test 4

1. To reach Simonville, the traveller needs to drive with extreme caution along the _____ curves of the mountain road that climbs _____ to the summit.

A. serpentine - steeply

B. jagged - steadily

C. gentle - precipitously

D. shady - steadily

E. hair-raising - languidly

2. The cricket match seemed _____ to our guests; they were used to watching sports in which the action is over in a couple of hours at the most.

A. unintelligible

B. inconsequential

C. interminable

D. implausible

E. evanescent

3. Our present accountant is most _____ ; unlike the previous _____ incumbent, he has never made a mistake in all the years that he has worked for the firm.

A. unorthodox - heretical

B. dependable - assiduous

C. punctilious - painstaking

D. asinine - diligent

E. meticulous - unreliable

4. The refugee's poor grasp of English is hardly an _____ problem; she can attend classes and improve within a matter of months.

A. implausible

B. insuperable

C. inconsequential

D. evocative

E. injudicious

5. We appreciated his _____ summary of the situation; he wasted no words yet delineated his position most _____ .

A. comprehensive, inadequately

B. succinct, direfully

C. cogent, persuasively

D. verbose, concisely

E. grandiloquent, eloquently

6. His musical tastes are certainly _____ ; he has recordings ranging from classical piano performances to rock concerts, jazz and even Chinese opera.

A. antediluvian

B. eclectic

C. harmonious

D. sonorous

E. dazzling

7. I cannot conclude this preface without _____ that an early and untimely death should have prevented Persius from giving a more finished appearance to his works.

A. rejoicing

B. lamenting

C. affirming

D. commenting

E. mentioning

8. Before his marriage the Duke had led an austere existence and now regarded the affectionate, somewhat _____ behavior of his young wife as simply _____ .

A. restrained - despicable

B. childish - elevating

C. playful - sublime

D. frivolous - puerile

E. unpleasant - delightful

9. Wilson _____ that human beings inherit a tendency to feel an affinity and awe for other living things, in the same way that we are _____ to be inquisitive or to protect our young at all costs.

A. argues - encouraged

B. maintains - trained

C. contends - predisposed

D. fears - taught

E. demurs - genetically programmed

10. The pond was a place of reek and corruption, of _____ smells and of oxygen-starved fish breathing through laboring gills.

A. fragrant

B. evocative

C. dolorous

D. resonant

E. fetid

11. There have been great _____ since his time, but few others have felt so strongly the _____ of human existence.

A. pessimists - futility

B. skeptics - exaltation

C. heretics - sagacity

D. optimists - tremulousness

E. cynics - joy

12. While war has never been absent from the _____ of man, there have been periods in History which appear remarkably _____ .

A. archives - ambivalent

B. posterity - serene

C. mind - desultory

D. annals - pacific

E. life - belligerent

PART ONE
Comprehension
Passage

PART TWO
Error Identification
Test

PART THREE
Sentence
completion Test

PART FOUR
Paragraph
Improvement Test

PART FIVE
Mock Test

Answers and Explanations

Test 4

1. A
由於司機需要非常小心，道路必須是危險的。因此，曲線可能是蛇形的(serpentine)，道路可能會陡峭地爬升(steeply)。
(serpentine=高度彎曲；jagged=邊緣粗糙；precipitously=陡峭；languidly=以輕鬆的方式)

2. C
分號後的部分提供了線索。它表示，他們習慣於觀看講求速度的東西，因此板球比賽似乎interminable(永無止境)。除了「evanescent(曇花一現)」這個詞意味著短暫的並且不符合這個意義之外，其他的詞都不是關於時間因素的。
(inconsequential=不重要；implausible=難以置信)

3. E
分號後面的「unlike」表示這兩個人是對立的。由於目前的會計師從未犯過錯，他是可靠的、謹慎的、細緻的，而前一個人是不可靠的(unreliable)。因此，第二條橫線上的選項會修正第一張的用詞，我們會選擇「一絲不苟」(meticulous)。
(unorthodox=異端的=反對傳統或正統的；punctilious=痛苦的=細緻的=勤奮的；asinine =愚蠢的)

4. B
由於難民可以迅速改善，她的英語不好並不是一個大問題。它可以克服，因此不是一個不可克服(insuperable)的問題。
(implausible=不可信的；insuperable=難以克服的、不可克服的；inconsequential=不重要的；evocative=令人回味的；injudicious=不明智的)

5. C
由於分號後的句子部分表明他「wasted no words」，所以他可以簡潔明了。因為我們讚賞他的總結，所以我們還需要第二條橫線上的正面消息。因此，我們可以刪除帶有「direfully」的選項(因具「可怕」的意思，較負面)，並選擇有「persuasively(說服力)」。
(succinct=簡短且表達良好；direfully=可怕；cogent=清楚、有說服力；verbose=使用太多詞語、冗長；concise=短而重要；grandiloquent=自命不凡；eloquent=雄辯滔滔、能言善辯)

6. B
句子的第二部分在分號前解釋或放大部分。由於他的錄音範圍廣泛，他的口味可以被描述為廣泛(eclectic)。
(eclectic=從不同來源獲取東西、不拘一格；antediluvian=舊式；sonorous=鏗鏘有力)

7. B
這句話表明，Persius的早逝使他無法正確地完成他的工作，因此，撰寫序言的作者可能會後悔(lament)這一事實。(注意：句子的語氣很重要；積極的話不符合這個意義。如果不是「should」一詞，「commenting」或「mentioning」也許適合——你不能說你「comment that something should prevent someone」。)
(lament=表示遺憾)

8. D
妻子的行為被描述為「affectionate(情深)」，所以我們需要一個類似重量的詞來表達第一條空線的位置上。
因此我們消除了「restrained(克制)」和「unpleasant(不愉快)」。此外，Duke被形容為「austere(嚴峻)」，不太可能對深情的行為作出許可。所以從餘下的選項中，我們選擇了負面詞「puerile」。
(frivolous=非嚴肅；sublime=威懾；puerile=幼稚)

9. C
「in the same way」表示我們正在尋找平行的想法。
因此，如果我們「inherit」了某種傾向，那麼同樣，我們會繼承另一種傾向。「predisposed」或「genetically　programmed」都適合，但由於後者與「demurs」意味著猶豫或拒絕，故令這是不合適的。「Contends」意味著「爭辯」是一個更好的選擇。

10. E
「reek and corruption」這個詞告訴我們，池塘里充滿了腐爛。因此，不好的氣味或惡臭(fetid)是最好的選擇。
(evocative=令人想起；dolorous=悲傷；resonant=迴響)

PART ONE
Comprehension Passage

PART TWO
Interpretation
Test

PART THREE
Sentence
completion Test

PART FOUR
Paragraph
Interpretation Test

PART FIVE
Mock Exam

11. A

句子結構表明，無論人的品質如何，在他身上都比其他人強。所以，對於兩條橫線上都需要填上帶有負面意思的字詞。

因此，我們可以採取「pessimists(悲觀主義者，一個想像最壞的人)，因為他會感覺到人類存在的「futility(無用性)」。

(futility=無效、毫無意義；skeptics=懷疑的人；exaltation=喜悅；heretics=異教徒、反對正統說話的人；sagacity=智慧；tremulousness=猶豫；cynics=憤世嫉俗者、不相信人性善良的人)

12. D

「While」在這裡表示相反的東西如下。所以，雖然戰爭從來沒有完全消失，但有一段時期看起來非常和平。因此，「serene(寧靜)」或「pacific(和平)」可能適合第二條橫線，但在「serene」的選項中，附帶了「posterity」，這意味著「後代」，所以不適合。

(archives=文件或存儲的記錄；ambivalent=模糊；desultory=斷斷續續；annals=記錄或編年史；pacific=和平；belligerent =侵略)

Test 5

1. The crew of the air balloon _____ the sand bags to help the balloon rise over the hill.

A. capsized

B. jettisoned

C. salvaged

D. augmented

E. enumerated

2. We were not fooled by his _____ arguments; his plan was obviously _____ .

A. cogent - brilliant

B. hackneyed - banal

C. convoluted - labyrinthine

D. specious - untenable

E. lucid – intelligible

3. Hawkins is _____ in his field; no other contemporary scientist commands the same respect.

A. disparaged

B. ignominious

C. obsolete

D. anachronistic

E. preeminent

4. The model paraded in front of the celebrities with _____ ; it was impossible to tell that this was her first assignment.

A. panache

B. opprobrium

C. shame

D. trepidation

E. terror

5. The term lead pencil is a _____ ; pencils are filled with graphite not lead.

A. misnomer

B. misdemeanor

C. peccadillo

D. euphemism

E. metaphor

6. The _____ weather forced us to stay indoors.

A. enticing

B. glorious

C. restorative

D. inclement

E. congenial

7. It will be hard to _____ Leonid now that you have so _____ him.

A. pacify - soothed

B. mollify - incensed

C. antagonize - irritated

D. anger - ruffled

E. subdue - subjugated

8. The lectures on quantum physics were invariably _____ ; the lecturer _____ his ill-prepared material in a manner guaranteed to send even the most ardent student to sleep.

A. stimulating - delivered

B. pedestrian - enthused about

C. soporific - droned

D. scintillating - intoned

E. arcane – marshaled

9. Elinor _____ to counteract her negative feelings, but only suc-ceeded in _____ them.

A. tried - allaying

B. hoped - mitigating

C. desired - ameliorating

D. hesitated - deprecating

E. endeavoured - intensifying

10. She was roundly condemned for her _____ ; she betrayed the woman to whom she owed her success.

A. truculence

B. perfidy

C. serendipity

D. pragmatism

E. discernment

11. The progress of the disease is _____ ; it spreads stealthily without any symptoms in the early stages.

A. dramatic

B. acute

C. blatant

D. insidious

E. inexorable

Answers and Explanations

Test 5

1. B
如果船員想要氣球上升，船員將不得不扔掉沙袋。因此，我們需要的這個詞：
「jettisoned(指被拋棄、被拋出)」。
(capsized=沉沒；salvaged=獲救；augmented=增加；enumerated =計數)

2. D
句子表明「we were not fooled(我們沒有被他的論點愚弄)」。因此，我們需要將
否定詞放在第一條橫線上，而「specious(似是而非)」是最好的選擇，因為這個
詞意味「虛假」或「欺騙」，至於「untenable」則是指「不能得到支持」，它
的搭檔詞也很有道理。
(Cogent=明確的；lucid=清晰的；hackneyed=平庸的=愚蠢的和非原創
的；convoluted=複雜的、迷宮式的、複雜且涉及的；intelligible=可以理解)

3. E
分號後的部分告訴我們Hawkins受到高度尊重。因此，唯一合適的詞是「preem-
inent(卓越)」。(其他選項都有負面的意思)
(disparaged=批評；ignominious=可恥；obsolete=過時；anachronistic=不合
時宜)

4. A
句子的第二部分告訴我們，這是她的第一項任務，但這是不可能的。因此，由於
她不是新手，所以她必須表現得很好(以自信的方式)。
(opopbrium =羞恥；trepidation =恐懼)

5. A
用這個名字叫鉛筆顯然犯上某種錯誤，因為句子告訴我們這種鉛筆裡充滿了
graphite(石墨)。因此，選用「misnomer(用詞不當)」作為答案是合適的，因為
它意味著「錯誤的名稱」。
(misdemeanor=不當行為；peccadillo=未成年人罪行；euphemism=用禮貌詞
語來掩蓋某些不良內容的詞彙；metaphor=隱喻)

6. D
由於天氣迫使我們留在室內，所以答案是某種消極詞。因此，唯一合適的詞是
「inclement(惡意)」。
(enticing=引誘；restorative=活力；congenial=適合、友善)

7. B

句子結構意味著需要對立面。(It will be hard to do something now that something else has happened.)

所以，B是最佳選擇。由於現在你已經激怒了(incensed)他，所以你很難緩和(mollify)他。

(incensed=惹怒；ruffled=惱怒；subjugated=壓制)

8. C

分號後的句子部分告訴我們：講座「guaranteed to send even the most ardent student to sleep(保證讓最熱心的學生入睡)」。這告訴我們這個講座很無聊或者睡覺。所以，「pedestrian」行人(dull沉悶)或「soporific」催眠(「sleep-inducing」誘發睡眠)本身會較適合。

但是我們消除了「行人」，因為它的搭檔詞「enthused about(熱衷於)」並不符合這個意義。

(droned=用單調的聲音說話；scintillating=閃閃發光；intoned=敘述)

9. E

「But」表示需要對立面。因此，這句話意味著她試圖克服消極情緒，但只能使它們變得更糟。因此，我們選擇「endeavoured(嘗試)」和= intensifying(強化、加強)」。

(allaying= mitigating=減輕嚴重性；ameliorating=改善；deprecating=批評)

10. B

「betrayed」一詞是線索。由於她背叛了某人，她將因為背叛而遭到譴責。Perfidy是最好的選擇，因為它意味著背叛。

(truculence=固執；serendipity=幸運巧合；pragmatism=實用性；discernment=判斷)

11. D

「stealthily(隱身)」這個詞是最佳線索。由於該疾病悄悄傳播，「insidious」(隱匿性)這個詞是恰當的。

(acute=尖銳的、突然的；blatant=明顯的；insidious=陰險的、以緩慢而有害的方式進行)

PART

FOUR

Paragraph
Improvement

IV. Paragraph Improvement

In this section, two draft passages are cited. For each passage, questions are set to test candidates' skills in improving the draft. The focus of the questions is on writing skills, not power of understanding.

Examples:

The sentences below are parts of the early draft of two passages, some parts of which may have to be rewritten. Read the passages and choose the best answer to the question.

1. Which of the following versions of sentence (8) provides the best link between sentences (7) and (9), reproduced below?

(7) It may be that a few of the products we have described are not available in some countries. (8) But it is possible to place an order via the Internet. (9) They will be dealt with promptly and efficiently.

A. Furthermore, it is possible to order via the Internet.

B. The Internet can be used in such circumstances.

C. Orders can, however, be placed via the Internet.

D. Sentence (8) as it is now. No change needed.

Answer : C

2. Which of the following is the best revision of sentence (3), reproduced below?

(3) Mistakenly believing that smoking is a sign of maturity those in authority must act today to protect our citizens of tomorrow.

A. It is a mistake to believe that smoking is a sign of maturity. Those in authority must act today to protect our citizens of tomorrow.

B. It is a mistake to believe that smoking is a sign of maturity, those in authority must act today to protect our citizens of tomorrow.

C. Mistaken in their belief that smoking is a sign of maturity those in authority must act today to protect our citizens of tomorrow.

D. Those in authority should act today. Our citizens of today are mistakenly believing that smoking is a sign of maturity. They must be protected

Answer : A

Test 1

(1) One of the problems for scientists has always been determining the age of the fossils and other materials they find underground. (2) One method of wide usage is carbon-14 dating. (3) Carbon-14 dating is pretty straightforward. (4) Carbon-14 is a harmless radioactive material found in the Earth's atmosphere. (5) Animals and plants absorb carbon-14 into their tissues while they are alive. (6) After they die, however, they cease to absorb carbon-14, and the carbon-14 in their tissues begins to decay. (7) The process of decay that makes carbon-14 dating possible. (8) Scientists who study radioactive material like carbon-14 have developed the term "half-life". (9) The "half-life" of a substance is the amount of time it takes for half of the atoms in a sample to decay. (10) Different radioactive materials have half-lives ranging from seconds to thousands of years. (11) Carbon-14 has a half-life of 5730 years.

(12) A scientist named Willard Libby was the first person to use carbon-14 dating, in 1949. (13) Libby used the concept of "half-life" to determine that plant and animal remains containing half the amounts of carbon-14 expected in a living specimen were 5730 years old. (14) Those remains containing less than half the expected amounts were older; the ones that contained more than half were younger.

1. In context, which of the following is the best version of the under-lined portion of sentence 2 (reproduced below)?

One method of wide usage is carbon-14 dating.

A. (as it is now)
B. One method has been widely used, it
C. One widely used method
D. One method the use of which has been wide

2. In context, which of the following is the best version of the under-lined portion of sentences 3 and 4 (reproduced below)?

Carbon-14 dating is pretty straightforward. Carbon-14 is a harmless radioactive material found in the Earth's atmosphere.

A. (as it is now)
B. is pretty straightforward. It
C. is pretty straightforward, because carbon-14
D. is pretty straightforward, carbon-14

3. In context, which of the following sentences would be best to insert between sentence 12 and sentence 13?

A. Libby was preoccupied with the problem of harmful radioactive waste.

B. His mentor, Dr. Frank McLean, had been the first geologist to em-ploy the carbon-14 method.

C. Libby realized that the plant and animal remains he was studying contained traces of carbon-14.

D. Libby believed that the concept of "half-life" was fundamentally flawed.

4. The main purpose of paragraph 2 is to

A. present a memorable example
B. summarize previous information
C. explain a puzzling contradiction
D. introduce an important concept

5. In context, what is the best way to deal with sentence 7 (reproduced below)?

The process of decay that makes carbon-14 dating possible.

A. Delete "the process of"
B. Delete "that"
C. Place a comma before "that"
D. Insert "is the thing" before "that"

6. Which of the following is the best way to phrase the underlined portion of sentence 14 (reproduced below)?

Those remains containing less than half the expected amounts were older; the ones that contained more than half were younger.

A. (as it is now)
B. those containing
C. the ones with a containment of
D. the remaining ones with

PART ONE
Comprehension
Passage

PART TWO
Cloze Procedure
Test

PART THREE
Sentence
Completion Test

PART FOUR
Paragraph
Improvement Test

PART FIVE
Mock Paper

Answers and Explanations

Test 1

1. C
選項C利用compound adjective(複合形容詞)「widely used」，簡而精地描繪了「method」。
選項A：用subordinate clause (附屬子句)「of wide usage」來修改「method」，會令句意變得含糊不清。
選項B：不令人滿意，因為會出現「逗號拼接」問題。
選項D：寫法過於囉嗦。

2. A
選項A將兩個完整、獨立的想法，分成兩個語法正確的句子。
選項B：第二句的主語是「carbon-14」，但第二句的開首卻用上代詞(pronoun)「It」，會誤導考生認為該主語是「carbon-14 dating」。
選項C：建立了這兩句之間不合邏輯的因果關係。「carbon-14 dating」的本質與「carbon-14」的無害性其實並無關聯。
選項D：句中用了逗號連接兩個獨立的子句，而沒有用適當的標點符號或連接詞。

3. C
選項C確定了導致Libby使用「carbon-14 dating」的原因。
選項A：不令人滿意，Libby被形容為working with無害放射性物質「carbon-14」
選項B：與第12句中的説法相矛盾，即Libby是第一個使用該方法的人。
選項D的説法並不令人滿意，因為Libby在不改變它的情況下使用「half-life」的概念。

4. D
本段介紹的「half-life」概念有助於解釋「carbon-14 dating」。
選項A：儘管第2段提供了一個材料的half-life（carbon-14）的例子，但該段的主要目的是解釋這個概念。
選項B：不令人滿意，因為第2段不涉及前述有關「carbon-14」的信息。
選項C：沒有試圖解釋第2段中的矛盾地方。

5. B

刪除「that」會令句子更見清晰、正確和簡潔。

選項A：不能令人滿意，因為它重複了原文的錯誤。

選項C：不令人滿意，因為添加逗號並不能解決句子的根本缺陷。

選項D：不令人滿意，因為這種加法令句子顯得囉嗦。

6. B

「those containing」的使用與前面的「those remains containing」在結構上相同或類似。

選項A：不令人滿意，因為雖然在技術上不正確，但「the ones that contained」違反了句子中「those remains containing」所確立的模式。

選項C：不令人滿意，因為它會造成囉嗦和模糊的句子。

選項D：不能令人滿意，因為它重複了原文的錯誤。

Test 2

(1) Shoppers in the United States beware—there's a new way to buy groceries, and it's coming to a store near you. (2) As recently as 1999, only 6% of U.S. supermarkets had self-checkout lines. (3) By 2003 the number had risen to 38%. (4) And half of those supermarkets in the survey without self-checkout said that they planned to add the service. (5) Hardware giant Home Depot was among the first major retailers to use these machines. (6) I remember when I first took the self-checkout plunge. (7) It was a large chain supermarket near my office. (8) One day I noticed four large electronic units taking up prime space in a prominent corner of the store, beckoning shoppers with the promise of shorter lines. (9) I was attracted, not fearful. (10) It's so easy to let the cashier scan the groceries, check the coupons, take the money—was I up to the challenge of do-it-yourself? (11) I saw myself all elbows and thumbs, fumbling with my groceries and wallet as I battled the flashing computer screen. (12) Circumstances were driving me to face my fears. (13) When I at last approached the head of my line, the cashier announced it was time for her break.

1. Which sentence should be deleted from the essay because it contains unrelated information?

A. Sentence 2

B. Sentence 5

C. Sentence 7

D. Sentence 8

2. All of the following strategies are used by the writer of the passage EXCEPT

A. statistical evidence

B. imaginative description

C. humorous example

D. direct quotation

3. In context, which of the following is the best phrase to insert at the beginning of sentence 12?

A. But as it turned out,

B. In addition,

C. The result was that

D. In most cases,

4. In context, which of the following is the best version of the under-lined portion of sentences 6 and 7 (reproduced below)?

I remember when I first took the self-checkout plunge. It was a large chain supermarket near my office.

A. (as it is now)

B. plunge. It happened that there was

C. plunge. It happened at

D. plunge, it happened to be at

5. The best way to describe the relationship of paragraph 2 to para-graph 1 is that paragraph 2

A. uses an anecdote to illustrate the facts presented in paragraph 1

B. provides facts to support a claim made in paragraph 1

C. tells a story that contradicts a theory offered in paragraph 1

D. uses statistics to back up a point made in paragraph 1

6. In context, which of the following is the best version of sentence 9 (reproduced below)?

I was attracted, not fearful.

A. (as it is now)

B. I was attracted and fearful.

C. I was attracted, because I was fearful.

D. I was attracted, yet fearful.

Answers and Explanations

Test 2

1. B
該段落討論超市雜貨店購物的自助結賬購物模式，作者在這篇文章的其他地方並沒有提到五金店。
選項A：並不令人滿意，因為第2句和第3句一起提供了第1句中提出的主張的統計證據。
選項C：不令人滿意，因為第7句確定了自助結賬的位置。
選項D：不令人滿意，因為句子8給出了使用機器的吸引力。

2. D
該段的作者沒有引用其他任何發言者的話。
選項A：不令人滿意，因為句子2和3提供統計證據。
選項B：不令人滿意，因為第11句提供了富想象力的描述。
選項C：不令人滿意的，因為作者的經驗是幽默的：他認為讓收銀員掃瞄他的雜貨比使用機器更容易，事實上它變得更加困難。

3. A
第12句通過暗示最終發生的事情——作者使用機器來對付第11句的焦慮。
選項B：不令人滿意，因為句子12標誌著前一句的語調和意義發生了變化；它不是一個簡單的「附加」。
選項C：不令人滿意，因為作者沒有建議他認為所出現的情況，是他焦慮的結果。
選項D：不令人滿意，因為作者不是討論「大多數情況」。

4. C
第7句清楚而簡明地標明了「plunge」的位置。
選項A：不令人滿意，因為它令令意思變得含糊：句子7中的「it」是指「super-market」還是「plunge」？
選項B：不令人滿意，因為寫法冗長且難以處理。
選項D：不令人滿意，因為它只用逗號。將兩個獨立的子句連接起來。

5. A
第2段的軼事說明了第1段中引入的增加自助結賬的事實。
選項B：不令人滿意，因為第2段中的相關信息並不是最好的描述為「事實」。
選項C：不令人滿意，因第2段中的相關故事支持第1段給出的信息。
選項D：不令人滿意，因為第2段不包含統計數字。

6. D
句子作為第8句和第10句之間的過渡，第8句強調了機器的吸引力，第10句強調
了作者的恐懼。這句話正確地表明了兩者之間的對比。
選項A：不令人滿意，因為第9句後面的信息表明作者對使用機器感到焦慮。
選項B：不令人滿意，因為雖然作者被證明既吸引又害怕，但第10句強調了作者
的恐懼。這個轉向需要用比「and」更強的詞來介紹。
選項C：不令人滿意，因為作者的吸引力和恐懼之間沒有因果關係。

Test 3

(1) Bizarre and mysterious events are things that happen all the time. (2) We have all had experiences that seemed supernatural, like unexpectedly meeting a long-lost friend when it was the day after dreaming about that friend. (3) Some people refer to these types of events as miracles. (4) Is there any scientific explanation for miracles?

(5) Professor Littlewood defined a miracle as an event with a probability of one in a million. (6) It is unlikely that you would ever see a miracle, right? (7) Well, Littlewood's Law of Miracles states that, for most people, miracles happen roughly once a month. (8) The proof of Littlewood's law is very simple. (9) He started by estimating that most people are active - at work or school, interacting with other people - at least 8 hours each day. (10) During this period, we see and hear things almost constantly; Littlewood estimated one new event per second. (11) These events add up fast because 1 per second for 8 hours equals 30,000 per day, or one million in one month.

(12) So if people are exposed to one million events per month, and a miracle is a one-in-a-million event, then we all experience a miracle once a month.

1. What is the best way to deal with sentence 1 (reproduced below)?

Bizarre and mysterious events are things that happen all the time.

A. Insert "to us" after "time"

B. Delete "and mysterious"

C. Change "are" to "could be"

D. Delete "are things that"

2. In context, what is the best version of the underlined portion of sentence 2 (reproduced below)?

We have all had experiences that seemed supernatural, like unexpectedly meeting a long-lost friend when it was the day after dreaming about that friend.

A. (as it is now)

B. friend the day after

C. friend because it was the day after

D. friend; it was the day after

3. In context, which of the following is the best phrase to insert at the beginning of sentence 6?

A. In any case,

B. Littlewood claimed that

C. Thus I believe

D. At that rate,

4. Which of the following sentences would be most logical to insert before sentence 5?

A. A math professor named Littlewood has tried to explain miracles mathematically.

B. Some people believe that miracles cannot be explained by science.

C. Professor Littlewood taught mathematics at Cambridge University in England.

D. If we all paid closer attention to mathematical principles, perhaps we could explain miracles.

5. In context, which is the best version of the underlined portion of sentence 11 (reproduced below)?

These events add up fast because 1 per second for 8 hours equals 30,000 per day, or one million in one month.

A. (as it is now)

B. add up too fast for 1 per second for 8 hours to equal

C. add up fast: 1 per second for 8 hours equals

D. add up to 1 per second for 8 hours because it equals

6. Sentence 12 in the passage is best described as

A. introducing a new topic

B. summarizing previous information

C. presenting a personal opinion

D. providing an additional example

Answers and Explanations

Test 3

1. D
「事件是事物」是多餘的。
選項A：不理想，因為該句子提出了一個要點；沒有必要指定作者正在討論「us」。
選項B：「bizarre」和「mysterious」代表了兩類不同的事件：「mysterious」事件不一定是「bizarre」事件。
選項C：不令人滿意，因為作者正在對神秘事件的流行，作簡單的陳述性陳述；這句話並沒有通過用「could be」這個短語限定陳述來改進。

2. B
選項B簡潔地表明了「who」和「when」。
選項A：不令人滿意，因為它包含了「when it was the day after」這個重複的短語。
選項C：不令人滿意，因為它引入了「dream」和「meeting」之間不合邏輯的因果關係。
選項D：不令人滿意，因為兩個獨立的子句在邏輯上並不相關。

3. D
「rate」一詞在邏輯上將這個句子連接到前面的句子。
選項A不令人滿意，因為「in any case」這個短語表明：第6句中的主張與Littlewood的調查結果無關。事實上，這兩者直接相關。
選項B：不令人滿意，因為Littlewood實際上卻反其道而行之。
選項C：不令人滿意，因為作者從不直接討論他或她自己的信仰。

4. A
Littlewood及其調查結果在本段開頭部分進行了適當介紹。這句話也使句子4和句子5之間有了明確的過渡。
選項B：不令人滿意，因為它沒有正確引入該段的主題Littlewood。
選項C：不令人滿意，因為它提供了關於Littlewood的不必要的信息，並提供了有關Littlewood的詳細信息，而不是他的工作介紹。
選項D：並不令人滿意，因為它沒有為Littlewood的討論提供必要的轉換。

5. C

冒號後的信息恰當地說明了前述權利要求的真相。

選項A：不令人滿意，因為它在索賠與其說明之間造成了不合邏輯的因果關係。

選項B：不令人滿意，因為該段落並不表示事件的發生率「too fast」。

選項D：不令人滿意，因為它在上下文中沒有意義。

6. B

句子12通過精確地總結前面介紹的信息(「the number of events」和「the rate of miracles」兩者)來結束這段話。

選項A：不令人滿意，因為在這句話中沒有引入新的主題。

選項C：不令人滿意，因為沒有任何跡象表明這句話陳述了作者的個人意見。

選項D：不令人滿意，因為這句話沒有提供新的例子。

Test 4

(1) Easter Island is a small island in the Pacific Ocean, over 2,000 miles off the coast. (2) Its distance from the coast of South America results in it being one of the most remote places in the world. (3) Easter Island is famous for its enormous stone statues of human heads and torsos. (4) Some of these statues are 70 feet tall and weigh over 200 tons.

(5) Scholars have puzzled over the origins of these statues, they are over 300 years old. (6) Archaeologists have located the stone quarry where the statues were built, but many of them are now located miles away. (7) How were such large objects moved? (8) Long thick ropes could have been made from tree bark and tied around the statues. (9) They could then have been dragged along the ground on sleds made of logs.

(10) This explanation seems to make sense. (11) European explorers arriving on Easter Island in the 1700s found no large trees or bushes anywhere on the island. (12) Ethnographer Kathleen Routledge visited Easter Island in 1914. (13) Where did the islanders get their ropes and sleds? (14) The answer lies in botany and its recent discoveries: Easter Island was once home to the world's largest type of palm tree.

1. In context, which is the best version of the underlined portion of sentences 1 and 2 (reproduced below)?

Easter Island is a small island in the Pacific Ocean, over 2,000 miles off the coast. Its distance from the coast of South America results in it being one of the most remote places in the world.

A. off the coast of South America. This is because it is

B. off the coast of South America. It is considered

C. off the coast. It is thus

D. off the coast. As far as South America goes, it is

2. Which sentence should be deleted from the essay because it contains unrelated information?

A. Sentence 14

B. Sentence 6

C. Sentence 9

D. Sentence 12

3. In context, which of the following revisions is necessary in sentence 5 (reproduced below)?

Scholars have puzzled over the origins of these statues, they are over 300 years old.

A. Insert "that" before "they"

B. Delete "they are"

C. Change "they" to "which"

D. Change "they" to "the statues"

4. Which of the following sentences would be most logical to insert before sentence 8?

A. The key seems to have been trees.

B. The islanders lacked sufficient organization.

C. Other groups utilized stone tools.

D. The only explanation is a supernatural one.

5. In context, which of the following is the best version of the under-lined portion of sentences 10 and 11 (reproduced below)?

This explanation seems to make sense. European explorers arriving on Easter Island in the 1700s found no large trees or bushes any-where on the island.

A. (as it is now)

B. seems to make sense, but European explorers

C. seems to make sense, and European explorers

D. seems to make sense, because European explorers

6. In context, what is the best version of the underlined portion of sentence 14 (reproduced below)?

The answer lies in botany and its recent discoveries: Easter Island was once home to the world's largest type of palm tree.

A. (as it is now)

B. To study botany recently is to discover the answer

C. A team of botanists recently uncovered the answer

D. A recent botanical discovery was the answer

Answers and Explanations

Test 4

1. B
選項B通過清晰、簡潔地呈現有關「Easter Island」位置的信息，來避免原始文件的錯誤。
選項A：不令人滿意，因為它不合邏輯地在 Easter Island 的位置和該位置的描述之間產生因果關係。
選項C：不令人滿意，因為它沒有提及「South America」，因此遺漏了重要的信息。
選項D：不令人滿意，因為它包含了一個模糊而無用的短語：「as far as South America goes」。

2. D
文中沒有提到Kathleen Routledge對Easter Island的訪問。它也不支持該段落中提出的任何觀點。
選項A：不令人滿意，因為句子14回答了上一句中提出的問題。
選項B：不令人滿意，因為第6句提供了關於雕像的建造和當前位置的關鍵信息。
選項C：不令人滿意，因為第9句解釋了雕像運輸過程中的一個關鍵點。

3. C
該建議的修訂通過修改具有從屬條款的「statues」來解決協調問題（用逗號連接的兩個獨立的子句）。
選項A：不令人滿意，因為「that they are over 300 years old」並沒有適當修改「statues」
選項B：不令人滿意，因為「over 300 years old」不是一個合適的從屬條款。
選項D：不令人滿意，因為修訂後的句子仍然將兩個獨立的子句連接在一起並用逗號連接。

4. A
「The key seems to have been trees」這句話有助於回答前面句子中提出的問題。另外，句子8和句子9都顯示了樹木如何被用來解決雕像的運輸問題。
選項B：不令人滿意，因為在這句話中提出的主張，與下一句提出的解釋相矛盾。
選項C：不令人滿意，因為這些信息與Easter Island上運送雕像的問題無關。
選項D：不令人滿意，因為隨後提出的解釋是自然而合理的。

PART ONE
Comprehension
Passage

PART TWO
Error Identification
Test

PART THREE
Sentence
Correction Test

PART FOUR
Paragraph
Improvement Test

PART FIVE
Mock Paper

5. B

歐洲探險家發現的證據，暫時引起了對擬議解釋的質疑。

選項A：不令人滿意，因為第二句中給出的證據與第一句相矛盾，而不是支持它。

選項C：不令人滿意，因為「and」直接連接了兩個相互矛盾的句子。

選項D：不令人滿意，因為它建立了不合邏輯的因果關係。探索者的證據並不導致解釋的意義；實際上，這破壞了解釋的邏輯。

6. C

選項C通過清晰簡潔地傳達關於植物學家發現的信息來避免原始錯誤。

選項A：不令人滿意，因為冒號前面的信息僅與冒號後面的信息間接相關。

選項B：不令人滿意，因為它以誇張的語言誇大其辭。

選項D：不能令人滿意，因為它含糊而囉嗦，並沒有清楚地介紹冒號後面的信息。

Test 5

(1) In the history of science, there are some discoveries that nobody can forget, one of them is when Copernicus proved that the Earth revolves around the Sun. (2) The scientists who made these groundbreaking discoveries must be among the smartest people of all time. (3) It's amazing that they could even come up with the ideas for their discoveries. (4) Much less go out and prove them. (5) But even the world's smartest people aren't perfect.

(6) Before Copernicus came along, people believed that the sun and the other planets circled around the Earth. (7) The models people used to show that the Earth was the center of the universe had stood the test of time: they were 1500 years old! (8) It's hard to believe anyone could overturn 1500 years of scientific teaching. (9) I know I couldn't. (10) But astronomers today can see that Copernicus's actual calculations were wrong. (11) He was correct in general, but he had trouble with the details. (12) The same thing is true of John Dalton, who founded modern atomic theory. (13) The fact that these scientists are so celebrated, despite their flaws, just shows how groundbreaking their new ideas actually were. (14) Dalton's ideas were incredibly powerful and influential, but he used the wrong mathematical formulas.

1. Which of the following is the best way to revise sentence 1 (reproduced below)?

In the history of science, there are some discoveries that nobody can forget, one of them is when Copernicus proved that the Earth revolves around the Sun.

A. Change "are" to "were"

B. Insert "and" between "forget" and "one"

C. Change "one of them is" to "such as"

D. Delete "the history of"

2. Of the following, which is the best way to revise and combine the underlined portions of sentences 3 and 4 (reproduced below)?

It's amazing that they could even come up with the ideas for their discoveries. Much less go out and prove them.

A. for their discoveries; and much less

B. for their discoveries; much less

C. for their discoveries and much less

D. for their discoveries, much less

3. Which of the following sentences would be most logical to insert before sentence 6?

A. Take Copernicus, for example.

B. Copernicus was a Polish astronomer.

C. Many scientists led unhappy lives.

D. Famous scientists, politicians, and military leaders have all made mistakes.

4. In context, which is the best version of the underlined portion of sentence 7 (reproduced below)?

The models people used to show that the Earth was the center of the universe had stood the test of time: they were 1500 years old!

A. (As it is now)
B. test of time, they were
C. test; their time was
D. test of time because they were

5. What should be done with sentence 9 (reproduced below)?

I know I couldn't.

A. Leave it as it is
B. Delete it
C. Insert "Even if I were a brilliant astronomer," at the beginning.
D. Insert "that" between "know" and "I".

6. What is the best way to deal with sentence 14 (reproduced below)?

Dalton's ideas were incredibly powerful and influential, but he used the wrong mathematical formulas.

A. Leave it as it is
B. Delete it
C. Insert "groundbreaking" before "ideas"
D. Switch it with sentence 13

PART ONE
Comprehension
Passage

PART TWO
Error Identification
Test

PART THREE
Sentence
Completion Test

PART FOUR
Paragraph
Improvement Test

PART FIVE
Mock Paper

Answers and Explanations

Test 5

1. C
將「one of them is」改為「such as」使得該短語遵從逗號而非獨立,這使得逗號的使用可接受。
選項A:不令人滿意,因為這樣的改變只會影響一個動詞;「can」和「is」也必須改為在過去式中創造連貫的句子。
選項B:不令人滿意,因為它糾正了原文的問題,但卻導致了一個尷尬而囉唆的句子。引入關於Copernicus信息的方法要比用「and one of them is」這個短語好得多。
選項D:不令人滿意,因為它不能糾正原來的問題:兩個獨立的想法只加入一個逗號作連接。

2. D
「Much less go out and prove them」是一個獨立的短語,應該將逗號加入獨立的短語「It's amazing…their discoveries」。
選項A:不令人滿意,因為分號後面的內容是相關的。分號應該用於連接兩個獨立的子句。
選項B:不令人滿意,因為它加入了一個從屬條款和一個帶有分號的獨立條款。分號應該用於連接兩個獨立的子句。
選項C:不令人滿意,因為依賴性短語「much less go out and prove them」增加了重點,應該用逗號加以引出,而不是用「and」。

3. A
Copernicus被視為一個不完美天才的「example」,因此將句子與前一句連接起來,而「Take Copernicus」一詞則正確地指出接下來的幾句話將進一步討論Copernicus。
選項B:不令人滿意,因為雖然它提供了第二段的邏輯引導,但它不能與前面的句子(句子5)直接相關。
選項C:不令人滿意,因為雖然它鬆散地延續了第5句的想法,但它處理的是一個問題,在這段內文的其他任何地方都沒有涉及:unhappiness。
選項D:並不令人滿意,因為該段不討論政治家或軍事領導人,只討論科學家。

4. A
在這句話中，冒號是正確的標點符號，因為「they were 1500 years old」解釋了為什麼人們可能會說「models…had stood the test of time」。
選項B：不令人滿意，因為它只用逗號加入了兩個獨立的想法。
選項C：不令人滿意，因為它沒有具體說明模型「stood」的「test」。此外，「test of time」是一個不應該被打破的慣用語。
選項D：不合格，因為它不合邏輯。這些模型由於年代久遠而未經受時間考驗；他們經受了時間的考驗，因為他們很老。

5. B
應刪除第9句，因為它會以與作者無關的陳述中斷流程。
選項A：並不令人滿意，因為第9句是一種不必要和令人分心的個人反思。
選項C：不能令人滿意，因為它不能糾正原文的問題：句子9是一篇不屬於論文的個人反思。
選項D：不令人滿意，因為第9句對該段落的主題和作者之間進行了無關的比較。

6. D
這兩個句子應該轉換，因為句子14支持第12句中關於Dalton的說法，句子13對兩位科學家提出了一個結論。
選項A：不令人滿意，因為它不合邏輯。第13句是指多個科學家的缺陷（「their flaws」），但第14句第一次提到Dalton的缺陷。
選項B：不令人滿意，因為Dalton的缺陷必須解釋為Dalton-Copernicus比較有意義。
選項C：不令人滿意，因為它是多餘的：「groundbreaking」的想法在定義上是「powerful and influential」。

Mock Paper I

1. Comprehension

This section aims to test candidates' ability to comprehend a written text. A prose passage of non-technical background is cited. Candidates are required to exercise skills in deciding on the gist, identifying main points, drawing inferences, distinguishing facts from opinion, interpreting figurative language, etc.

PART ONE
Comprehension
Passage

PART TWO
Error Identification
Test

PART THREE
Sentence
Completion Test

PART FOUR
Paragraph
Improvement Test

PART FIVE
Mock Paper

Passage A

In the 16th century, an age of great marine and terrestrial exploration, Ferdinand Magellan led the first expedition to sail around the world. As a young Portuguese noble, he served the king of Portugal, but he became involved in the quagmire of political intrigue at court and lost the king's favor. After he was dismissed from service by the king of Portugal, he offered to serve the future Emperor Charles V of Spain.

A papal decree of 1493 had assigned all land in the New World west of 50 degrees W longitude to Spain and all the land east of that line to Portugal. Magellan offered to prove that the East Indies fell under Spanish authority. On September 20, 1519, Magellan set sail from Spain with five ships. More than a year later, one of these ships was exploring the topography of South America in search of a water route across the continent. This ship sank, but the remaining four ships searched along the southern peninsula of South America. Finally they found the passage they sought near 50 degrees S latitude. Magellan named this passage the Strait of All Saints, but today it is known as the Strait of Magellan.

One ship deserted while in this passage and returned to Spain, so fewer sailors were privileged to gaze at that first panorama of the Pacific Ocean. Those who remained crossed the meridian now known as the International Date Line in the early spring of 1521 after 98 days on the Pacific Ocean. During those long days at sea, many of Magellan's men died of starvation and disease.

Later, Magellan became involved in an insular conflict in the Philippines and was killed in a tribal battle. Only one ship and 17 sailors under the command of the Basque navigator Elcano survived to complete the westward journey to Spain and thus prove once and for all that the world is round, with no precipice at the edge.

1. The 16th century was an age of great _____ exploration.

A. cosmic

B. land

C. mental

D. common man

E. None of the above

2. Magellan lost the favor of the king of Portugal when he became involved in a political _____ .

A. entanglement

B. discussion

C. negotiation

D. problem

E. None of the above

PART ONE
Comprehension
Passage

PART TWO
Proofreading
Test

PART THREE
Sentence
Completion Test

PART FOUR
Paragraph
Improvement Test

PART FIVE
Mock Paper

3. The Pope divided New World lands between Spain and Portugal according to their location on one side or the other of an imaginary geographical line 50 degrees west of Greenwich that extends in a _____ direction.

A. north and south

B. crosswise

C. easterly

D. south east

E. north and west

4. One of Magellan's ships explored the _____ of South America for a passage across the continent.

A. coastline

B. mountain range

C. physical features

D. islands

E. None of the above

5. Four of the ships sought a passage along a southern _____ .

A. coast

B. inland

C. body of land with water on three sides

D. border

E. Answer not available

6. The passage was found near 50 degrees S of _____.

A. Greenwich
B. The equator
C. Spain
D. Portugal
E. Madrid

7. In the spring of 1521, the ships crossed the _____ now called the International Date Line.

A. imaginary circle passing through the poles
B. imaginary line parallel to the equator
C. area
D. land mass
E. Answer not available

公務員入職　**英文運用熱門試題王**

PART ONE
Comprehension
Passage

PART TWO
Error Identification
Test

PART THREE
Sentence
completion Test

PART FOUR
Paragraph
Improvement Test

PART FIVE
Mock Paper

Answers and Explanations

Passage A

1. B
「Terrestrial」指土地。這裡並沒有選項提供「marine」(海洋)的同義詞，例如 nautical/naval/water/seagoing，沒有其他選擇可以與海洋或陸地相匹配。

2. A
「Quagmire」的字面意思是「沼澤」或「沼澤」，並且形象地説是一種「難以 逃脱」的情況；「entanglement」是一個同義詞，比其他選項更形似。

3. A
「Longitudes」(經度)是虛構的南、北向之地理線。「Latitudes」(緯度)則向 東、西運行。其他選項在方向上不等於經度或緯度。

4. C
「Topography」意味陸塊(land mass)的物理特徵，而不是「coastline(海岸線) 」(選項A)、「mountain range (山脈)」(選項B)或「islands (島嶼)」(選項D)。

5. C
「A peninsula」(半島)是一塊通過地峽與大陸相連，並通向海洋的土地，其三面 被水包圍。「A peninsula」不是「coast(海岸)」(選項A)；不是內陸(B)，且也不 是邊界(D)。

6. B
文中所指「通道」(The passage)發現於北緯50度附近。緯度是相對於赤道或中 心虛線水平測量的，在北極和南極之間等距離。經度是垂直測量的。「Green-wich(格林威治)」(選項A)作為全球標準採用的零度經度的位置是錯誤的，並且 從未在通道中被命名，至於「Spain(西班牙)」(選項C)、「Portugal(葡萄牙)」(選項D)和「Madrid(馬德里)」(選項E)也並不正確。

7. A
「Meridians」是與南、北兩極相交的想象地理圈。平行於赤道(B)的假想線是緯 度。「The International Date Line」是特定子午線，而不是「area(區域)」(選項 C)。它不是「land mass(陸地塊)」(選項D)，因為它跨越了水和土地。

Passage B

The Trojan War is one of the most famous wars in history. It is well known for the 10-year duration, for the heroism of a number of legendary characters, and for the Trojan horse. What may not be familiar, however, is the story of how the war began.

According to Greek myth, the strife between the Trojans and the Greeks started at the wedding of Peleus, King of Thessaly, and Thetis, a sea nymph. All of the gods and goddesses had been invited to the wedding celebration in Troy except Eris, goddess of discord. She had been omitted from the guest list because her presence always embroiled mortals and immortals alike in conflict.

To take revenge on those who had slighted her, Eris decided to cause a skirmish. Into the middle of the banquet hall, she threw a golden apple marked "for the most beautiful". All of the goddesses began to haggle over who should possess it. The gods and goddesses reached a stalemate when the choice was narrowed to Hera, Athena, and Aphrodite. Someone was needed to settle the controversy by picking a winner. The job eventually fell to Paris, son of King Priam of Troy, who was said to be a good judge of beauty. Paris did not have an easy job. Each goddess, eager to win the golden apple, tried aggressively to bribe him.

"I'll grant you vast kingdoms to rule," promised Hera. "Vast kingdoms are nothing in comparison with my gift," contradicted Athena. "Choose me and I'll see that you win victory and fame in war." Aphrodite out-did her adversaries, however. She won the golden apple by offering Helen, daughter of Zeus and the most beautiful mortal in the land, to Paris. Paris, anxious to claim Helen, set off for Sparta in Greece.

Although Paris learned that Helen was married, he nevertheless accepted the hospitality of her husband, King Menelaus of Sparta. Therefore, Menelaus was outraged for a number of reasons when Paris departed, taking Helen and much of the king's wealth back to Troy. Menelaus collected his loyal forces and set sail for Troy to begin the war to reclaim Helen.

1. Eris was known for _____ both mortals and immortals.

A. scheming against
B. creating conflict amongst
C. feeling hostile toward
D. ignoring
E. comforting

2. Each goddess tried _____ to bribe Paris.

A. boldly
B. effectively
C. secretly
D. carefully
E. Answer not available

3. Athena _____ Hera, promising Paris victory and fame in war.

A. disregarded the statement of

B. defeated

C. agreed with

D. restated the statement of

E. questioned the statement of

PART ONE
Comprehension
Passage

PART TWO
Error Identification
Test

PART THREE
Sentence
Completion Test

PART FOUR
Paragraph
Improvement Test

PART FIVE
Mock Paper

Answers and Explanations

Passage B

1. B
文中説，Eris的存在是「goddess of discord」(不和諧的女神) :「總是把凡人和不朽者捲入衝突。」將他們捲入衝突會在他們之間造成衝突。它並不意味著「scheming against(詭計)」(選項A)他們，「scheming against(對他們有敵意)」(選項C)，「ignoring(無視)」(選項D)他們，或「carefully(安慰)」(選項E)他們。

2. A
「Aggressively」意味著大膽。這並不意味著「effectively(有效)」(選項B)或成功、「secretly(秘密)」(選項C)或「carefully(小心)」(選項D)。

3. A
「Contradicted」意味著Athena無視Hera的聲明並對此提出異議或反駁。這並不意味著她「defeated(擊敗了)」(選項B)她的陳述，「agreed with(同意)」(選項C)它，「restated the statement of (重申)」(選項D)或「questioned the statement of(質疑)」(選項E)它。

II. Error Identification

Knowledge on use of the language is tested through identification of language errors which may be lexical, grammatical or stylistic.

Example :

The sentence below may contain a language error. Identify the part (underlined and lettered) that contains the error or choose 'E No error' where the sentence does not contain an error.

Irrespective for the outcome of the probe, the whole sorry affair has already cast a shadow over this man's hitherto unblemished record as a loyal servant to his country.

A. Irrespective for
B. sorry
C. cast
D. hitherto
E. No error

Answer : A

1. Mary is not sure if she should buy the new computer now or wait until he receives her next bonus.

A. if

B. should

C. or

D. receives

E. No error

2. Ivy scored poorly on the exam, which is not surprising since she did not prepare adequately.

A. poorly

B. which

C. since

D. adequately

E. No error

3. I prefer John to any hairdresser I have visited in the past because he has such a good understanding of his clients' needs.

A. prefer

B. any

C. have visited

D. clients'

E. No error

4. The archivist <u>had</u> not <u>only</u> a deep <u>interest</u> but also a clear under-standing of the historical documents <u>in</u> the museum.

A. had

B. only

C. interest

D. in

E. No error

5. With skill and <u>surprising</u> gentleness the fireman <u>managed to</u> lower the injured cat <u>down</u> from the <u>top of</u> the tree.

A. surprising

B. managed to

C. down

D. top of

E. No error

6. Was the woman <u>who</u> you think you <u>saw</u> leaving the building <u>wearing</u> a <u>nurse's</u> uniform?

A. who

B. saw

C. wearing

D. nurse's

E. No error

7. As he <u>held open</u> the door for her, she could not ignore the look on his face, a look that <u>aggravated</u> her self-consciousness as they <u>proceeded</u> along the street.

A. held open

B. ,

C. aggravated

D. proceeded

E. No error

8. Many people genuinely want <u>to be</u> fitter, but <u>few</u> have the tenacity <u>for sticking</u> to a suitable <u>regime</u> of diet and exercise.

A. to be

B. few

C. for sticking

D. regime

E. No error

9. Safety precautions and emergency exits, matters of great <u>concern for</u> builders of commercial establishments, <u>are</u> often <u>overlooked</u> when <u>designing</u> a new home.

A. concern for

B. are

C. overlooked

D. designing

E. No error

10. From the time he <u>took up</u> his new position <u>as</u> head of the department, he <u>has been</u> <u>concerned about</u> the legitimacy of his appointment.

A. took up

B. as

C. has been

D. concerned about

E. No error

PART ONE
Comprehension
Passage

PART TWO
Error Identification
Test

PART THREE
Sentence
Completion Test

PART FOUR
Paragraph
Improvement Test

PART FIVE
Mock Paper

Answers and Explanations

1. A
將「if」改為「whether」。

2. B
「Which」在這裡的用法不正確，因為「is not surprising」並不是指「Which」前面的那個詞(exam)。

3. B
由於John看來是位理髮師，故我們需要在「any」後加插「other」。

4. C
我們需要説「interest in」。

5. C
「Down」在本句是多餘的。

6. A
將「who」改為「whom」。

7. E
這句沒有錯誤。

8. C
正確的寫法為「tenacity to stick」。

9. D
我們不知道誰在「designing」新的家。

10. E
這句沒有錯誤。

III. Sentence Completion

In this section, candidates are required to fill in the blanks with the best options given. The questions focus on grammatical use.

Example :

Complete the following sentence by choosing the best answer from the options given.

This market research company claims to predict in advance _____ by conducting exit polls of selected voters.

A. the results of an election will be
B. the results will be of an election
C. what results will be of an election
D. what the results of an election will be
E. what will the results of an election be

Answer : D

PART ONE
Comprehension
Passage

PART TWO
Error Identification
Test

PART THREE
Sentence
Completion Test

PART FOUR
Paragraph
Improvement Test

PART FIVE

Mock Paper

1. The candidate _____ when asked why he had left his last job; he did not want to admit that he had been _____ .

A. demurred - promoted

B. confided - banned

C. dissembled - dismissed

D. rejoiced - wrong

E. hesitated - lauded

2. Tennyson was a well-loved poet; no other poet since has been so _____ .

A. lionized

B. attacked

C. decried

D. poetical

E. abhorred

3. The parliamentary session degenerated into _____ with politicians hurling _____ at each other and refusing to come to order.

A. mayhem - banter

B. disarray - pleasantries

C. tranquility - invectives

D. chaos - aphorisms

E. anarchy - insults

4. The admiral _____ his order to attack when he saw the white flag raised by the enemy sailors; he was relieved that he could bring an end to the _____ .

A. reiterated - hostilities

B. countermanded - fighting

C. commandeered - truce

D. renounced - hiatus

E. confirmed - aggression

5. In a fit of _____ she threw out the valuable statue simply because it had belonged to her ex-husband.

A. pique

B. goodwill

C. contrition

D. pedantry

E. prudence

6. Many 17th century buildings that are still in existence have been so _____ by successive owners that the original layout is no longer _____ .

A. preserved - visible

B. modified - apparent

C. decimated - enshrouded

D. salvaged - required

E. neglected - appropriate

7. Since ancient times sculpture has been considered the _____ of men; women sculptors have, until recently, consistently met with _____ .

A. right - acceptance

B. domain - approbation

C. domicile - ridicule

D. realm - condolence

E. prerogative - opposition

8. _____ action at this time would be inadvisable; we have not yet accumulated sufficient expertise to warrant anything other than a _____ approach.

A. precipitate - cautious

B. hesitant - wary

C. vacillating - circuitous

D. decisive - firm

E. ponderous - direct

9. Many biologists have attempted to _____ the conditions on earth before life evolved in order to answer questions about the _____ of biological molecules.

A. mimic - fitness

B. standardize - shapes

C. replicate - reactions

D. simulate - origin

E. ameliorate - evolution

10. Harding was unable to _____ the results of the survey; although entirely unexpected, the figures were obtained by a market research firm with an _____ reputation.

A. accept - peerless

B. discount - impeccable

C. fault - mediocre

D. counter - unenviable

E. believe - fine

PART ONE
Comprehension
Passage

PART TWO
Error Identification
Test

PART THREE
Sentence
Completion Test

PART FOUR
Paragraph
Improvement Test

PART FIVE
Mock Paper

Answers and Explanations

1. C

句子的第二部分告訴我們他不想承認某件事。因此，我們可以為第一條橫線選擇一個字詞，意思是猶豫(hesitates)或避免(avoids): demurred, dissembled, or hesitated。然而，他不會介意承認他被提升或lauded(讚美)，所以我們選擇了dismissed。

(demurred=被猶豫或拒絕；dissembled=避免説出真相)

2. A

「Well-loved(深受喜愛)」告訴我們，我們想要的詞是「loved(被愛)」或類似的東西。Lionized是合適的，因為它意味著「treated like a celebrity(像名人一樣對待)」。

(decried=批評；abhorred =討厭)

3. E

由於政治家「refused to come to order(拒絕接受訂單)」，會議必定已經演化成無序。因此，除了tranquility(安寧)之外，任何第一個空白字都可能是合適的。

接下來我們看看他們在互相「hurling(兜售)」什麼。這需要像侮辱或「invectives(謾罵)」，但我們可以排除褻瀆，因為它是安寧的合作夥伴。

(mayhem=混亂和混亂 disarray=chaos=anarchy；tranquility=和平；banter=好玩的談話；pleasantries=笑話；invectives=侮辱；aphorisms=眾所周知的説法)

4. B

句子的意義告訴我們，當他看到白旗（投降的信號）時，他會放鬆結束戰鬥，並取消他的命令。

因此，我們選擇説他拒絕了(countermanded)他的命令，並且結束了戰鬥。

(reiterated=重複；countermanded=取消；commandeered=被徵用、佔有；truce=和平協議；hiatus=中斷)

5. A

她因為屬於前夫而拋出了一尊寶貴的雕像。因此，她一定是出於惡意或惡意行事。因此，我們選擇「pique(怨恨)」。(注意：「pique」用作動詞意味著「刺激好奇心」。)

(contrition=懺悔；pedantry=堅持狹隘的學習點；prudence=謹慎)

6. B

「no longer(不再)」表示狀態的改變。因此原始佈局將不再可見或明顯。如果佈局不再明顯，則建築物必須「modified(進行修改)」。

7. E

句子的兩半表明男性和女性雕塑家的治療存在差異。如果雕塑被認為是男性的特權，那麼女性雕塑家就會遇到反對意見。請注意，儘管「ridicule(嘲笑)」可能適合第二條橫線，但其合作夥伴「domicile」的意思是「家」，並且不適合。

8. A

問問自己：如果我們沒有積累專業知識，哪種行為是不明智的。由於「precipitate(沉澱)」意味著匆忙，這似乎是最好的選擇。也沒有專業知識，只有「謹慎(cautious)」或謹慎的做法才是合適的。

(vacillating=搖擺、猶豫；circuitous=間接；ponderous=緩慢而沉重)

9. D

最好的線索是我們正在處理「conditions on earth before life evolved(生命演化前的地球條件)」。因此，生物學家可能試圖「replicate(複製)」，「mimic(模仿)」或模擬這些條件，以找出生物分子是如何進化的。因為我們正在考慮進化，所以第二個空白位置的最好詞就是「evolution(進化)」或起源。我們選擇後者是因為它與我們的第一個空白選項配對。

10. B

如果市場研究公司的聲譽很好，即使是出乎意料的，Harding也不能拒絕調查結果。因此，我們選擇「discount(折扣、忽略)」、「impeccable(無可挑剔)」。

(peerless=無平等；mediocre=平均；unenviable=不被嫉妒)

PART ONE Comprehension Corpus?

PART TWO Error Identification Test

PART THREE Sentence Completion Test

PART FOUR Paragraph Improvement Test

PART FIVE
Mock Paper

IV. Paragraph Improvement Passage A

(1) The early history of astronomy was full of misunderstandings. (2) Some of them were funny, it's like the controversy of the "canali" on Mars. (3) In the late 1800's an Italian astronomer named Giovanni Schiaparell studied Mars. (4) He had a high-powered telescope that hused to look at Mars. (5) Schiaparelli thought he saw channels criss-crossing the planet's surface. (6) He was intrigued: perhaps these channels were evidence that Marhad great flowing rivers like the Earth. (7) Schiaparelli made charts of the surface of Mars and labeled it with the Italian word "canali".

(8) Unfortunately, "canali" can be translated into English as either "channels" or "canals." (9) Channels and canals are two different things because channels are form naturally by water, while canals are constructed by people(10) Some people translated "canali" as "canals", word began to spread that the lines Schiaparelli saw through his telescope were actually canals that had been built by intelligent beings. (11) One of them was an amateur astronomer named Percival Lowell. (12) He wrote a serie of best-selling books. (13) In these books Lowell publicized the notion that these "canals" were built by Martian farmers who understood irrigation.

(14) In 1965 a U.S. spacecraft flying close to the surface of Mars sent back conclusive pictures. (15) There are no prominent channels anywhere on the planet. (16) Lowell and Schiaparelli saw what they wanted to see. (17) Lowell was wrong, of course, but so was Schiaparelli.

1. Which is the best version of the underlined portion of sentence 2 (reproduced below)?

Some of them were funny, it's like the controversy of the "canali" on Mars.

A. (as it is now)

B. funny; it's like

C. funny, like

D. as funny as

2. Which is the best way to combine sentences 3 and 4 (reproduced below)?

In the late 1800's an Italian astronomer named Giovanni Schiaparelli studied Mars. He had a high-powered telescope that he used to look at Mars.

A. In the late 1800's an Italian astronomer named Giovanni Schiaparelli studied Mars by a high-powered telescope.

B. In the late 1800's an Italian astronomer named Giovanni Schiaparelli studied Mars with a high-powered telescope that he used to look at Mars.

C. In the late 1800's an Italian astronomer named Giovanni Schiaparelli studied Mars, he had a high-powered telescope that he used.

D. In the late 1800's an Italian astronomer named Giovanni Schiaparelli used a high-powered telescope to study Mars.

3. Which word would be best to insert at the beginning of sentence 10 (reproduced below)?

Some people translated "canali" as "canals", word began to spread that the lines Schiaparelli saw through his telescope were actually canals that had been built by intelligent beings.

PART ONE
Comprehension
Passage

PART TWO
Error Identification
Test

PART THREE
Sentence
completion Test

PART FOUR
Paragraph
Improvement Test

PART FIVE

Mock Paper

A. Whereas

B. However

C. If

D. Because

4. What is the best version of the underlined portion of sentence 11 (reproduced below)?

One of them was an amateur astronomer named Percival Lowell.

A. (As it is now)

B. One of the most intelligent was

C. This idea was popularized by

D. The person who solved the problem was

5. What is the best way to combine sentences 12 and 13 (reproduced below)?

He wrote a series of bestselling books. In these books Lowell publicized the notion that these "canals" were built by Martian farmers who understood irrigation.

A. In a series of bestselling books, Lowell publicized the notion that these "canals" were built by Martian farmers who understood irrigation.

B. He wrote a series of books that was a bestseller and publicized the notion that these "canals" were built by Martian farmers who understood irrigation.

C. His books that were bestsellers publicized the notion that these "canals" were built by Martian farmers who understood irrigation.

D. In these books, which were bestsellers, Lowell publicized the notion that these "canals" were built by Martian farmers who understood irrigation.

Answers and Explanations

Passage A

1. C
這句話恰當地介紹了文段主題的爭議,這是一些有趣的誤解之一。
選項A:並不令人滿意,因為它只用一個逗號,連接兩個獨立的想法。
選項B:不合格,因為它不合邏輯;「it's」在這種情況下沒有意義。
選項D:是不令人滿意的,因為「as funny as」這句話的重點在於其他誤解,並期望它們將與「canali」爭論相比較。這樣的比較在文段中沒有發生。

2. D
這句話很好地連接句子3和4的想法,而沒有不必要地重複信息。
選項A:不令人滿意,因為寫Schiaparelli「studied Mars by」並不符合語言習慣——如果用「with」或「through」會更合適。
選項B:不令人滿意,因為它重複了有關不必要地研究/查看有關火星的信息。
選項C:不令人滿意,因為它只用逗號連接兩個完整的想法。

3. D
「Because」一詞恰當地表示了翻譯問題與對「canals」的誤解之間的關係。
選項A:並不令人滿意,因為「Whereas」表明,「canali」的錯譯和對建造「canali」的誤解是矛盾的觀點。
選項B:不令人滿意,因為「However」在這方面沒有意義。
選項C:不令人滿意,因為「If」表明對於人們是否以這種方式誤譯「canali」存在疑問。這段話的含義是,人們錯誤地翻譯了這個詞,導致對建造「canali」存在誤解。

4. C
選項C正確地表明了前一句中提出的觀點與天文學家Lowell之間的關係。
選項A:不令人滿意,因為代名詞「them」似乎指的是在火星上建造運河的「intelligent beings」,而Lowell顯然不是其中之一。
選項B:不令人滿意,因為在文章的其他地方沒有將Lowell描述為「intelligent」;事實上,他的理論被證明是基於一個簡單的誤解。
選項D:不令人滿意,因為根據該段落,Lowell創造的問題比他解決的還要多。

PART ONE
Comprehension
Basics

PART TWO
Error Identification
Test

PART THREE
Sentence
Continuation Test

PART FOUR
Paragraph
Improvement Test

PART FIVE

Mock Paper

5. A

選項A所得出的句子能夠保持原句(句子13)的有效結構，同時加入了句子12的重要信息。

選項B：不令人滿意，因為將一系列的書籍描述為「a bestseller」是不恰當的。

選項C：不令人滿意，因為在第11句之後，假如句子主語為「Lowell」的話，會較「his books」合適。

選項D：不令人滿意，因為它表明 Lowell 的書籍在文段之前提過 （「In these books」），但這才是他們第一次提到它。

Passage B

(1) Not many children leave elementary school and they have not heard of Pocahontas' heroic rescue of John Smith from her own people, the Powhatans. (2) Generations of Americans have learned the story of a courageous Indian princess who threw herself between the Virginia colonist and the clubs raised to end his life. (3) The captive himself reported the incident. (4) According to that report, Pocahontas held his head in her arms and laid her own upon his to save him from death.

(5) But can Smith's account be trusted? (6) Probably it cannot, say several historians interested in dispelling myths about Pocahontas. (7) According to these experts, in his eagerness to find patrons for future expeditions, Smith changed the facts in order to enhance his image. (8) Portraying himself as the object of a royal princess' devotion may have merely been a good public relations ploy. (9) Research into Powhatan culture suggests that what Smith described as an execution might have been merely a ritual display of strength. (10) Smith may have been a character in a drama in which even Pocahontas was playing a role.

(11) As ambassador from the Powhatans to the Jamestown settlers, Pocahontas headed off confrontations between mutually suspicious parties. (12) Later, after her marriage to colonist John Rolfe, Pocahontas traveled to England, where her diplomacy played a large part in gaining support for the Virginia Company.

PART ONE
Comprehension
Passage

PART TWO
Error identification
Test

PART THREE
Sentence
Completion Test

PART FOUR
Paragraph
Improvement Test

PART FIVE

Mock Paper

1. What is the best way to deal with sentence 1 (reproduced below)?

Not many children leave elementary school and they have not heard of Pocahontas' heroic rescue of John Smith from her own people, the Powhatans.

A. Leave it as it is.

B. Switch its position with that of sentence 2.

C. Change "leave" to "have left".

D. Change "and they have not heard" to "without having heard".

2. In context, which of the following is the best way to revise the underlined wording in order to combine sentences 3 and 4?

The captive himself reported the incident. According to that report, Pocahontas held his head in her arms and laid her own upon his to save him from death.

A. The captive himself reported the incident, according to which

B. Since then, the captive reported the incident, which said that

C. According to the captive's own report of the incident,

D. It seems that in the captive's report of the incident he says that

3. Which of the following phrases is the best to insert at the beginning of sentence 10 to link it to sentence 9?

A. Far from being in mortal danger,

B. If what he says is credible,

C. What grade school history never told you is this:

D. They were just performing a ritual, and

4. Which of the following best describes the relationship between sentences 9 and 10?

A. Sentence 10 concludes that the theory mentioned in sentence 9 is wrong.

B. Sentence 10 adds to information reported in sentence 9.

C. Sentence 10 provides an example to illustrate an idea presented in sentence 9.

D. Sentence 10 poses an argument that contradicts the point made in sentence 9.

5. Which of the following would be the best sentence to insert before sentence 11 to introduce the third paragraph?

A. It is crucial to consider the political successes as well as the shortcomings of Pocahontas.

B. The Pocahontas of legend is the most interesting, but the historical Pocahontas is more believable.

C. If legend has overemphasized the bravery of Pocahontas, it has underplayed her political talents.

D. To really know Pocahontas, we must get beyond myth and legend to the real facts about her private life.

PART ONE
Comprehension
Passage

PART TWO
Cloze comprehension
Test

PART THREE
Sentence
completion Test

PART FOUR
Passage
improvement Test

PART FIVE

Mock Paper

Answers and Explanations

Passage B

1. D

它恰當地解釋了大多數孩子在離開elementary　school前都聽過有關Pocahontas的故事。

選項A：不令人滿意，因為原來的句子連接了兩個主要思想——離開學校的孩子和Pocahontas的故事——但當中只有連詞「and」。因此，這句話並沒有提供這兩種觀點間關係的線索。

選項B：並不令人滿意，因為在講故事之前說出故事或事件中主要人物的名字是合乎邏輯的，而不是之後——特別是當名字很熟悉時。

選項C：不令人滿意，因為它重複了原文的錯誤，未能解釋孩子與故事之間的關係。

2. C

選項C產生的句子既能保留原文的意義，同時避免冗餘。

選項A：不令人滿意，因為它包含了一個不清晰的指涉：「which」似乎指的是事件本身，而不是該報告。

選項B：並不令人滿意，因為「which」一詞似乎是指事件，因為它在邏輯上只能提及事件報告。

選項D：不令人滿意，因為它使用不必要的短語「it seems」作關聯。

3. A

選項A通過解釋上一句中提到的「ritual　display」的無害性，來將句子10與文段其餘部分聯繫起來（因此澄清了Smith的賬戶和可能的事實間之對比）。

選項B：不令人滿意，因為第10句提出了一個挑戰Smith「life-or-death」的說法，這意味著Smith不是一個可靠的來源。

選項C：不令人滿意，因為插入的短語中斷了句子9中引入的「ritual display」與第10句中的解釋彼此之間的聯繫。

選項D：不令人滿意，因為使用「and」意味著「ritual」和「drama」是兩個不同的事件，而「drama」實際上是指「ritual display」。

4. B
第10句詳細說明了Smith在第9句中提出的真實情況。
選項A：並不令人滿意，因為第10條只支持第9條提出的要求。
選項C：不令人滿意，因為第10句中的信息不是「example」。相反，這是對Smith可能發生的事情的合理澄清。
選項D：不令人滿意，因為沒有任何關於句子10與句子9相矛盾。

5. C
第三段給出了Pocahontas晚年在政治上取得成功的兩個詳細例子。
選項A：不令人滿意，因為該段落沒有提到Pocahontas的任何缺點。
選項B：不令人滿意，因為關注歷史事實的可信度是奇怪和不必要的。
選項D：不令人滿意，因為第3段中的信息主要涉及Pocahontas的「public life」，而不是其「private life」。

Mock Paper II

Passage A

Conflict had existed between Spain and England since the 1570s. England wanted a share of the wealth that Spain had been taking from the lands it had claimed in the Americas.

Elizabeth I, Queen of England, encouraged her staunch admiral of the navy, Sir Francis Drake, to raid Spanish ships and towns. Though these raids were on a small scale, Drake achieved dramatic success, adding gold and silver to England's treasury and diminishing Spain's supremacy.

Religious differences also caused conflict between the two countries. Whereas Spain was Roman Catholic, most of England had become Protestant. King Philip II of Spain wanted to claim the throne and make England a Catholic country again. To satisfy his ambition and also to retaliate against England's theft of his gold and silver, King Philip began to build his fleet of warships, the Spanish Armada, in January 1586.

Philip intended his fleet to be indestructible. In addition to building new warships, he marshaled 130 sailing vessels of all types and recruited more than 19,000 robust soldiers and 8,000 sailors. Although some of his ships lacked guns and others lacked ammunition, Philip was convinced that his Armada could withstand any battle with England.

The martial Armada set sail from Lisbon, Portugal, on May 9, 1588, but bad weather forced it back to port. The voyage resumed on July 22 after the weather became more stable.

PART ONE
Comprehension Passage

PART TWO
Tutor Identification Test

PART THREE
Sentence Completion Test

PART FOUR
Paragraph Improvement Test

PART FIVE
Mock Paper

The Spanish fleet met the smaller, faster, and more maneuverable English ships in battle off the coast of Plymouth, England, first on July 31 and again on August 2. The two battles left Spain vulnerable, having lost several ships and with its ammunition depleted. On August 7, while the Armada lay at anchor on the French side of the Strait of Dover, England sent eight burning ships into the midst of the Spanish fleet to set it on fire. Blocked on one side, the Spanish ships could only drift away, their crews in panic and disorder. Before the Armada could regroup, the English attacked again on August 8.

Although the Spaniards made a valiant effort to fight back, the fleet suffered extensive damage. During the eight hours of battle, the Armada drifted perilously close to the rocky coastline. At the moment when it seemed that the Spanish ships would be driven onto the English shore, the wind shifted, and the Armada drifted out into the North Sea. The Spaniards recognized the superiority of the English fleet and returned home, defeated.

1. Sir Francis Drake added wealth to the treasury and diminished Spain's _____ .

A. unlimited power

B. unrestricted growth

C. territory

D. treaties

E. Answer not available in article

2. King Philip recruited many _____ soldiers and sailors.

A. warlike

B. strong

C. accomplished

D. timid

E. inexperienced

3. The _____ Armada set sail on May 9, 1588.

A. complete

B. warlike

C. independent

D. isolated

E. Answer not available

公務員入職　*英文運用熱門試題王*

4. The two battles left the Spanish fleet _____.

A. open to change
B. triumphant
C. open to attack
D. defeated
E. discouraged

5. The Armada was _____ on one side.

A. closed off
B. damaged
C. alone
D. circled
E. Answer not available in this article

Answers and Explanations

Passage A

1. A
「Supremacy(霸權)」意味著無限的力量，而不是「unrestricted growth(無限制的增長)」(選項B)。文章指出，Drake削弱了西班牙的霸權地位，但沒有具體提到其領土的縮小(選項C)。Drake的突襲壯大了英格蘭並減少了西班牙的力量；沒有提到消除任何條約(Treaties，選項D)。

2. B
「Robust」意味著健壯、強大。它並不意味著「warlike(好戰)」(選項A)、「accomplished(完成)」(選項C)或勝任，「timied(膽小)」(選項D)或恐懼，或「inexperienced(沒有經驗)」(選項E)。

3. B
「Martial」意味著好戰或與戰爭有關。這並不意味著「complete(完整)」(A)、「independent(獨立)」(選項C)或「isolated(孤立)」(選項D)。

4. C
「Vulnerable」意味著容易受到攻擊或容易受到傷害。它並不意味著「open to change(可以改變)」(選項A)或接受、「triumphant(勝利)」(選項B)或勝利、「defeated(擊敗)」(選項D)或毆打——他們首先容易受到攻擊，然後被擊敗或「discouraged(氣餒)」(選項E)。

5. A
文段表明艦隊在一側被「blocked(封鎖)」，即封閉而不是「damaged(損壞)」(選項B)(它被廣泛地損壞，而不是在一側)；「alone(單獨)」(選項C)或「circled(圈出)」(選項D)，即被包圍，兩者都不能僅在一側完成。

PART ONE
Comprehension
Passage

PART TWO
Word Meaning in
Test

PART THREE
Sentence
Completion Test

PART FOUR
Paragraph
Improvement Test

PART FIVE
Mock Paper

Passage B

The victory of the small Greek democracy of Athens over the mighty Persian Empire in 490 B.C. is one of the most famous events in history. Darius, king of the Persian Empire, was furious because Athens had interceded for the other Greek city-states in revolt against Persian domination. In anger the king sent an enormous army to defeat Athens. He thought it would take drastic steps to pacify the rebellious part of the empire.

Persia was ruled by one man. In Athens, however, all citizens helped to rule. Ennobled by this participation, Athenians were prepared to die for their city-state. Perhaps this was the secret of the remarkable victory at Marathon, which freed them from Persian rule. On their way to Marathon, the Persians tried to fool some Greek city-states by claiming to have come in peace. The frightened citizens of Delos refused to believe this. Not wanting to abet the conquest of Greece, they fled from their city and did not return until the Persians had left. They were wise, for the Persians next conquered the city of Eritrea and captured its people.

Tiny Athens stood alone against Persia. The Athenian people went to their sanctuaries. There they prayed for deliverance. They asked their gods to expedite their victory. The Athenians refurbished their weapons and moved to the plain of Marathon, where their little band would meet the Persians. At the last moment, soldiers from Plataea reinforced the Athenian troops.

The Athenian army attacked, and Greek citizens fought bravely. The power of the mighty Persians was offset by the love that the Athenians had for their city. Athenians defeated the Persians in both archery and hand combat. Greek soldiers seized Persian ships and burned them, and the Persians fled in terror. Herodotus, a famous historian, reports that 6,400 Persians died, compared to only 192 Athenians.

1. Athens had _____ the other Greek city-states against the Persians.

A. refused help to
B. intervened on behalf of
C. wanted to fight
D. given orders for all to fight
E. defeated

2. Darius took drastic steps to _____ the rebellious Athenians.

A. weaken
B. destroy
C. calm
D. irritate
E. Answer not available

3. Their participation _____ to the Athenians.

A. gave comfort

B. gave honor

C. gave strength

D. gave fear

E. gave hope

4. The people of Delos did not want to _____ the conquest of Greece.

A. end

B. encourage

C. think about

D. daydream about

E. Answer not available

5. The Athenians were _____ by some soldiers who arrived from Plataea.

A. welcomed

B. strengthened

C. held

D. captured

E. Answer not available

Answers and Explanations

Passage B

1. B
「Interceded for」代表干預，「not refused help to (不拒絕幫助)」(選項A)、
「wanted to fight(想打架/打仗)」(選項C)，給予命令全面打擊(選項D)，或打敗
對方(選項E)。

2. C
「Pacify」意味著安撫、冷靜或和平。它並不意味著「weaken(變弱)」(選項A)
、「destroy(摧毀)」(選項B)或「irritate(刺激)」(選項D)，即惹惱或激怒。

3. B
「Ennobled」意味著榮耀、高尚或崇高。這並不意味著「gave comfort(給予安
慰)」(選項A)或慰借，「gave strength(給予力量)」(選項C)，即強化或強化，
「gave fear(給予恐懼)」(選項D)或恐懼，或「gave hope(給予希望)」(選項E)或
鼓勵。

4. B
「abet(慫恿)」意味著通常在犯罪或做錯事情時實現、支持或鼓勵。這並不意味
著「end(結束)」(選項A)，「think about(想想)」(選項C)或「daydream about(
做白日夢)」(選項D)。

5. B
「Reinforced」意味著加強，「not welcomed」(選項A)，「held(持有)」(選項
C)或「captured(俘獲)」(選項D)。

PART ONE   PART TWO   PART THREE   PART FOUR   **PART FIVE**
Comprehension Error Identification   Sentence   Paragraph
   Passage   Test   Completion Test Improvement Test

Mock Paper

II. Error Identification

Knowledge on use of the language is tested through identification of language errors which may be lexical, grammatical or stylistic.

Example :

The sentence below may contain a language error. Identify the part (underlined and lettered) that contains the error or choose 'E No error' where the sentence does not contain an error.

Irrespective for the outcome of the probe, the whole sorry affair has already cast a shadow over this man's hitherto unblemished record as a loyal servant to his country.

A. Irrespective for
B. sorry
C. cast
D. hitherto
E. No error

Answer : A

1. The presence of strong feeling, the cause <u>of which</u> is not fully understood, always has the <u>effect</u> of making <u>we</u> human <u>beings</u> uneasy.

A. of which

B. effect

C. we

D. beings

E. No error

2. The new law is <u>too</u> stringent<u>; it</u> will <u>be</u> neither respected <u>or</u> obeyed.

A. too

B. ; it

C. be

D. or

E. No error

3. I do not wish <u>to make</u> a formal complaint, but I would have been <u>better pleased</u> if you <u>gave</u> the award to the person <u>who</u> best deserved it.

A. to make

B. better pleased

C. gave

D. who

E. No error

4. <u>After you have written</u> a definition in your vocabulary notebook, add a <u>few</u> ways to use the word <u>or</u> a sentence to illustrate <u>it's</u> meaning.

A. After you have written

B. few

C. or

D. it's

E. No error

5. The tribesmen made offerings <u>to placate</u> the gods, <u>whom</u>, they believed, <u>were</u> angry with <u>them</u>.

A. to placate

B. whom

C. were

D. them

E. No error

6. Many physicists initially <u>regarded</u> quantum theory <u>as</u> unnatural, absurd, <u>and</u> incompatible <u>to</u> common sense.

A. regarded

B. as

C. and

D. to

E. No error

7. <u>From ancient times</u>, sculpture <u>had been</u> considered the <u>prerogative</u> of men, and even now, in some parts of the world, women sculptors <u>face</u> hostility and suspicion.

A. From ancient times

B. had been

C. prerogative

D. face

E. No error

8. It is an old criticism of the medical profession that <u>they</u> have <u>considered</u> the symptoms and <u>causes of</u> disease without sufficient <u>reference to</u> the causes of health.

A. they

B. considered

C. causes of

D. reference to

E. No error

9. <u>According to</u> Hume, it is not logic and reasoning <u>that</u> <u>determine</u> our actions, <u>but</u> emotion.

A. According to

B. that

C. determine

D. but

E. No error

PART ONE
Comprehension
Passage

PART TWO
Error Identification
Test

PART THREE
Sentence
Completion Test

PART FOUR
Paragraph
Improvement Test

PART FIVE

Mock Paper

10. The ornate pillars and <u>life-size</u> statues that the magnate <u>has chosen</u> to adorn <u>his</u> swimming pool are like <u>a Greek temple</u>.

A. life-size
B. has chosen
C. his
D. a Greek temple
E. No error

Answers and Explanations

1. C
將「we」改為「us」。

2. D
將「or」改為「nor」。

3. C
將「gave」改為「had given」，因為前面的「子句」（clause）中的動詞是以「past conditional tense」的時式顯示。

4. D
不需要使用撇號。

5. B
「Who」是必需的，因它是動詞「were angry」的主語。

6. D
將「incompatible to」改為「incompatible with」。

7. B
不需要「過去完成式」(past perfect tense)。將「had been」改為現在完成式 (present perfect tense)「has been」。

8. A
由於代名詞(pronoun)沒有先行詞(antecedent)的，故要將「they」改為「doctor」。

9. E
這句沒有錯誤。

10. D
不正確的比較。應寫成「are like those of a Greek temple」。

PART ONE
Comprehension
Passage

PART TWO
Error Identification
Test

PART THREE
Sentence
Completion Test

PART FOUR
Paragraph
Improvement Test

PART FIVE
Mock Paper

III. Sentence Completion

In this section, candidates are required to fill in the blanks with the best options given. The questions focus on grammatical use.

Example :

Complete the following sentence by choosing the best answer from the options given.

This market research company claims to predict in advance _____ by conducting exit polls of selected voters.

A. the results of an election will be
B. the results will be of an election
C. what results will be of an election
D. what the results of an election will be
E. what will the results of an election be

Answer : D

1. His one vice was gluttony and so it is not surprising that as he aged he became increasingly _____ .

A. emaciated

B. despondent

C. corpulent

D. carping

E. lithe

2. Our once thriving High School Nature Club is now _____ ; the programs have had to be cancelled due to lack of support.

A. defunct

B. extant

C. resurgent

D. burgeoning

E. renovated

3. Having been chief accountant for so many years, Ms. George felt herself to be _____ and was unwilling to _____ control of the department after the merger.

A. slighted - truncate

B. irreplaceable - assume

C. insubordinate - retain

D. decisive - continue

E. indispensable - relinquish

4. Because Elaine's father was a field entomologist who trekked over the continent studying insect infestations, and insisted on taking his young family with him, Elaine and her brother had a(n) _____ childhood.

A. idyllic

B. itinerant

C. sedentary

D. propitious

E. equable

5. Frederica was _____ when her supervisor took only a _____ look at her essay over which she had taken so much care.

A. exultant - superficial

B. vexed - studious

C. disappointed - cursory

D. pleased - patronizing

E. relieved - perfunctory

6. When he was young he _____ ideas of becoming a doctor; however, he was _____ by his father who wanted him to join the family business.

A. harbored - backed

B. entertained - dissuaded

C. produced - critical

D. repudiated - deterred

E. eschewed - encouraged

7. Literary criticism has in recent years become increasingly _____ ; it is almost impossible for the non-literary person to understand its analyses.

A. abstruse

B. accessible

C. colloquial

D. wide-ranging

E. professional

8. The alchemists, though they are often supposed to have been _____ or confidence tricksters, were actually skilful technologists.

A. empiricists

B. polemicists

C. pragmatists

D. theorists

E. charlatans

9. Bullock carts and hand pumps seem _____ in a village whose skyline is dominated by telephone cables and satellite dishes.

A. Anachronisms

B. exigencies

C. diversions

D. provocations

E. portents

PART ONE
Comprehension
Passage

PART TWO
Error Identification
Test

PART THREE
Sentence
completion Test

PART FOUR
Paragraph
improvement Test

PART FIVE
Mock Paper

10. A _____ child, she was soon bored in class; she already knew more mathematics than her junior school teachers.

A. obdurate

B. querulous

C. precocious

D. recalcitrant

E. contemporary

Answers and Explanations

1. C
「Gluttony(暴食)」告訴我們，他很貪婪，吃得很多。所以，如果他變胖，這並不奇怪。「Corpulent」意味著肥胖。

2. A
「Once」後跟「Now」表示事情已經改變。會所以前的業績良好(it was thriving)，現在情況很糟糕。此外，這句話告訴我們這些節目已經被取消了，所以我們可以選擇「defunct(停止使用)」，這意味著「不再運作」。

3. E
經過總會計師之後，Ms. George可能會認為她是「irreplaceable(不可替代)」或「indispensable(不可或缺)」。如是這樣，她不會「放棄」(relinquish)控制權。

4. B
句子告訴我們Elaine和弟弟在這片大陸上徒步旅行。因此，她們一定有一個「itinerant childhood(流浪的童年)」。

5. C
遵循句子的含義。如果老師只是粗略地(cursory)看了一下她(指學生)極度關心的事情，她(該名學生)會感到「失望(disappointed)」的。

6. B
However表明需相反的想法，因此我們可說他想成為醫生卻被父阻(dissuaded)

7. A
分號後的句子部分放大了第一部分陳述的內容。因此，如果我們可以說人們不理解它，文學批評一定會變得很模糊、深奧、艱澀(abstruse)。

8. E
「Though」和「actually」表明需要相反的想法。這個空白必須是「confidence tricksters」的另一個詞，而技術專家則相反，因此他們是charlatans(騙子)。

9. A
這句指牛車在現代村莊似已過時，因此Anachronisms(時代錯誤)這詞是必要。

10. C
這句話告訴我們，她遠遠領先於她的班級。這表明需要「precocious(早熟)」這個詞，這意味著一年中有天賦或進步。

IV. Paragraph Improvement Passage A

(1) People today have placed emphasis on the kinds of work that others do, it is wrong. (2) Suppose a woman says she is a doctor. (3) Immediately everyone assumes that she is a wonderful person, as if doctors were incapable of doing wrong. (4) However, if you say you're a carpenter or mechanic, some people think that you're not as smart as a doctor or a lawyer. (5) Can't someone just want to do this because he or she loves the work?

(6) Also, who decided that the person who does your taxes is more important than the person who makes sure that your house is warm or that your car runs ? (7) I know firsthand how frustrating it can be. (8) They think of you only in terms of your job. (9) I used to clean houses in the summer because the money was good; but yet all the people whose houses I cleaned seemed to assume that because I was vacuuming their carpets I did not deserve their respect. (10) One woman came into the bathroom while I was scrubbing the tub. (11) She kept asking me if I had any questions. (12) Did she want me to ask whether to scrub the tub counter-clockwise instead of clockwise ? (13) Her attitude made me angry! (14) Once I read that the jobs people consider important have changed. (15) Carpenters used to be much more admired than doctors. (16) My point is, then, that who I want to be is much more important than what I want to be!

1. Of the following, which is the best way to phrase sentence 1 (reproduced below)?

People today have placed emphasis on the kinds of work that others do, it is wrong.

A. (As it is now)

B. People today place too much emphasis on the kinds of work that others do.

C. The wrong emphasis is being placed today on people and what kind of work they do.

D. The wrong kind of emphasis had been placed on the kinds of work others do today.

2. In context, which of the following is the best way to revise and combine the underlined portions of sentences 2 and 3 (reproduced below)?

Suppose a woman says she is a doctor. Immediately everyone assumes that she is a wonderful person, as if doctors were incapable of doing wrong.

A. Suppose a woman says she is a doctor, but immediately

B. If a woman says she is a doctor, for instance, immediately

C. When a woman says she is a doctor, however, immediately

D. Immediately, if they say, for example, she is a doctor,

3. In context, the phrase do this in sentence 5 would best be replaced by:

A. hold this particular opinion

B. resist temptation

C. ask someone for assistance

D. become a carpenter or a mechanic

4. Which of the following is the best way to revise and combine the underlined portions of sentences 7 and 8 (reproduced below)?

I know firsthand how frustrating it can be. They think of you only in terms of your job.

A. be; they--people, that is--think of you

B. be when they are thinking of one

C. be how people think of you

D. be when people think of you

5. In context, the phrase but yet in sentence 9 would best be replaced by:

A. incidentally,

B. however,

C. in fact,

D. in addition,

Answers and Explanations

Passage A

1. B
選項B：too much成功地在首子句中嵌入否定判斷，使第二個子句變得不必要。
選項A：不令人滿意，因為它使用了逗號不恰當地加入了兩個獨立的子句。
選項C：因單調性而不令人滿意。句子不需名詞「people」和代名詞「they」。
選項D：過去完成式時（had been placed）不適合描述今天發生動作。

2. B
在組合句子中，dependent clause（由「if」適當引入）陳述可能的條件，然後主要子句描述可能的結果。「For instance」表示情況說明了句子1中的陳述。
選項A：連接詞「but」通常引入對比而不是結果。
選項C：過渡詞「however」不恰當暗示一個對比想法會隨之而來。
選項D：不令人滿意，因為它使用了一個模糊的代詞，「they」。

3. D
在選項D的上下文中，動詞「become」比「do」更精確，名詞「carpenter」和「mechanic」比代詞「this」更具體。
選項A：抱負面思想的人應不會是熱愛自己工作的人所可能出現行為。
選項B：不令人滿意，因為它引入了不相關的想法。這段內文中沒有提到誘惑。
選項C：兩個子句之間關係變得不合邏輯。愛一個人工作不一定產生援助請求。

4. D
選項D中的subordinating conjunction「when」在兩個子句之間提供了適當的連接，名詞「people」代替了「they」這個含糊的代詞。
選項A：代詞「they」和片語「that is」是不需要的。
選項B：不令人滿意，因為它保留了含糊不清的代詞「they」，並引入了另一個不適當的代詞「one」（與後面的代詞「your」不一致）。
選項C：不令人滿意，因為「how」不是將第一個子句（以「be」結尾）與第二個子句聯繫起來的可接受過渡詞。

5. B
「however」表明工作的積極（good pay）和消極（lack of respect）間對比。
選項A：「incidentally」不表示對比。相反，它表明所遵循的信息不重要。
選項C：in fact沒為對比做準備。
選項D：「in addition」一詞不能引入對比。這實際意味第二個子句將繼續或強化前面提出的想法。

Passage B

(1) Aristotle was a great philosopher and scientist. (2) Aristotle lived in Greece over 2300 years ago. (3) Aristotle was extraordinarily curious about the world around him. (4) He was also a master at figuring out how things worked. (5) Aristotle passed it on to his pupil Theophrastus.

(6) Theophrastus was famous among his contemporaries as the co-founder of the Lyceum, a school in Greece, he is best known today as "the father of botany". (7) Botany is the branch of science dealing with plants.

(8) Two famous books he wrote were Natural History of Plants and Reasons for Vegetable Growth. (9) His books were translated from Greek into Latin in 1483-1800 years after he wrote them-they influenced thousands of readers.

(10) Theophrastus made accurate observations about all aspects of plant life, including plant structure, plant diseases, seed use, and medicinal properties. (11) He even described the complex process of plant reproduction correctly, hundreds of years before it was formally proven. (12) In 1694 Rudolph Jakob Camerarius used experiments to show how plants reproduced. (13) According to some accounts, Theophrastus did his research in a garden he maintained at his school which was called the Lyceum. (14) But Theophrastus also wrote about plants that grew only in other countries, which he heard about from returning soldiers. (15) By comparing these plants to plants he grew in his garden, Theophrastus established principles that are still true today.

1. Which of the following is the best version of the underlined portion of sentence 1 and sentence 2 (reproduced below)?

Aristotle was a great philosopher and scientist. Aristotle lived in Greece over 2300 years ago.

A. philosopher and a scientist, living

B. philosopher and scientist who lived

C. philosopher, and, as a scientist, lived

D. philosopher and scientist; Aristotle lived

2. What would best replace "it" in sentence 5?

A. that

B. them

C. these traits

D. the world

3. What word should be inserted between "Greece", and "he" in sentence 6 (reproduced below)?

Theophrastus was famous among his contemporaries as the co-founder of the Lyceum, a school in Greece, he is best known today as "the father of botany".

A. and

B. but

C. for

D. thus

4. Which sentence should be inserted between sentence 8 and sentence 9?

A. Theophrastus's ideas had a lasting impact.

B. Theophrastus's books were instantly successful.

C. The first book is still studied today in botany classes.

D. They challenged the conclusions of Aristotle.

5. Which revision appropriately shortens sentence 13 (reproduced below)?

According to some accounts, Theophrastus did his research in a garden he maintained at his school which was called the Lyceum.

A. Delete "his school which was called".

B. Delete "According to some accounts,".

C. Delete "in a garden he maintained".

D. Replace "According to some accounts" with "Therefore".

Answers and Explanations

Passage B

1. B
選項B恰當地使用了一個「relative clause(關係子句)」（「who lived...」）將第2句中關於Aristotle的信息，與第1句中的信息連接起來。
選項A：不令人滿意，因為它導致了一個尷尬的句子。
選項C：不令人滿意，因為它不必要地將Aristotle作為哲學家的想法，與Aristotle作為科學家的想法分開。根據這段內文，他在2300年前在希臘實踐了哲學和科學。
選項D：不令人滿意，因為結果的句子不必要地重複「Aristotle」。

2. C
第三句和第四句是指Aristotle擁有的品質或特質。第五句表明Aristotle把東西傳給了他的學生Theophrastus，而我們從上下文中知道Theophrastus擁有這兩種特質，所以說Aristotle將這些特質傳給了Theophrastus是有道理的。
選項A：不令人滿意，因為該段指出了Aristotle擁有的兩個特徵。斷定他只傳授其中一個給Theophrastus是不合邏輯的。
選項B：並不令人滿意，因為當Aristotle把兩種特徵傳給了Theophrastus時，這些特徵需要被確定為這樣；「them」在上下文中是模棱兩可的。
選項D：不令人滿意，因為說Aristotle把「the world」傳遞給他的學生是毫無意義的。

3. B
連接詞「but」恰當地表明了關於Theophrastus名聲的兩種對比陳述之間的關係。
選項A：不令人滿意，儘管「and」創造了一個正確的句子，但它在語境中與「but」不一樣有效。「and」這個詞表明，關於Theophrastus的名聲的兩個陳述是互補的事實，沒有任何對比。
選項C：不令人滿意，因為「for」一詞在上下文中沒有意義。
選項D：不令人滿意，因為「thus」這個詞意味著不存在的因果關係。

4. A
在選項A中，插入的句子在邏輯上連接到句子8（「ideas」明確地指代書籍的內容）和句子9（「lasting impact」由1483翻譯的成功顯示）並且支持在第一句的段落。
選項B：並不令人滿意，因為儘管Theophrastus的書可能立刻獲得成功（儘管他

PART ONE
Comprehension
Passage

PART TWO
Error Identification
Test

PART THREE
Sentence
Completion Test

PART FOUR
Paragraph
Improvement Test

PART FIVE

Mock Paper

在其他事物中同時代著名的事實表明了另外的事實），但這一成功並未在文章中提及，並且與上下文無關。

選項C：不令人滿意，因為插入的句子按時間順序排列順序。在討論Theophrastus在1400年代的相關性之後，今天討論Theophrastus的相關性會更有意義。

選項D：並不令人滿意，因為在Theophrastus的著作質疑Aristotle的結論的這段説法中，沒有任何支持。

5. A

結尾句子「…did his research in a garden he maintained at the Lyceum」由於讀者已經知道The Lyceum是Theophrastus學校的名字（第6句），所以這一修訂是合適和必要的。

選項B：並不令人滿意，因為「According to some accounts」，表明對所提供的信息存在一些懷疑。刪除這個重要的限定詞是不合適的。

選項C：不令人滿意，因為它消除了關於花園的重要事實並保持原始冗餘。

選項D：不令人滿意，因為「therefore」在上下文中沒有意義。

看得喜 放不低

創出喜閱新思維

書名	公務員入職 英文運用 熱門試題王（第三版）
ISBN	978-988-74807-2-3
定價	HK$118
出版日期	2021年5月
作者	Fong Sir
責任編輯	投考公務員系列編輯部
版面設計	陳沫
出版	文化會社有限公司
電郵	editor@culturecross.com
網址	www.culturecross.com
發行	香港聯合書刊物流有限公司
	地址：香港新界大埔汀麗路36號中華商務印刷大廈3樓
	電話：（852）2150 2100
	傳真：（852）2407 3062
台灣總經銷：	貿騰發賣股份有限公司
	電話：（02）822 75988

網上購買 請登入以下網址：

一本 My Book One
（www.mybookone.com.hk）

超閱網 Superbookcity
（www.mybookone.com.hk）

香港書城 Hong Kong Book City
（www.hkbookcity.com）